THE KYOTO PROTOCOL AND BEYOND

LEGAL AND POLICY CHALLENGES OF CLIMATE CHANGE

T.M.C. ASSER INSTITUUT
The Hague

THE KYOTO PROTOCOL AND BEYOND

LEGAL AND POLICY CHALLENGES OF CLIMATE CHANGE

edited by

W.TH. DOUMA, L. MASSAI AND M. MONTINI

T·M·C·ASSER PRESS

Published by T·M·C·ASSER PRESS
P.O. Box 16163, 2500 BD The Hague, The Netherlands
<www.asserpress.nl>

T·M·C·ASSER PRESS' English language books are distributed exclusively by:

Cambridge University Press, The Edinburgh Building, Shaftesbury Road,
Cambridge CB2 2RU, UK,
or
for customers in the USA, Canada and Mexico:
Cambridge University Press, 100 Brook Hill Drive, West Nyack, NY 10994-2133, USA
<www.cambridge.org>

ISBN 978-90-6704-972-6

**T.M.C. Asser Instituut – Institute for Private and Public International Law,
International Commercial Arbitration and European Law**

Institute Address: R.J. Schimmelpennincklaan 20-22, The Hague, The Netherlands; Mailing Address: P.O. Box 30461, 2500 GL The Hague, The Netherlands; Tel.: +3170 342 0300; Fax: +3170 3420 359; Internet: www.asser.nl.

Over forty years, the T.M.C. Asser Institute has developed into a leading scientific research institute in the field of international law. It covers private international law, public international law, including international humanitarian law, the law of the European Union, the law of international commercial arbitration and increasingly, also, international economic law, the law of international commerce and international sports law.

Conducting scientific research, either fundamental or applied, in the aforementioned domains, is the main activity of the Institute. In addition, the Institute organizes congresses and postgraduate courses, undertakes contract-research and operates its own publishing house.

Because of its inter-university background, the Institute often cooperates with Dutch law faculties as well as with various national and foreign institutions.

FOREWORD

I had the pleasure of participating at the two conferences which form the basis of this book: as a chairman at the 2007 The Hague Conference '*Tackling Climate Change – An appraisal of the Kyoto Protocol and options for the future*' and as a speaker at the 2006 Siena Conference '*The Kyoto Protocol and beyond: a legal perspective*'. I would like to thank my colleagues Wybe Douma, Leonardo Massai and Massimiliano Montini for those opportunities, and although I was, unfortunately, unable to contribute a paper to this book due to time constraints, I am glad to be able to say a few words on the issue by means of this foreword.

The timing of the two conferences was well chosen: the period between the Siena Conference (June 2006) and The Hague Conference (March 2007) encapsulated perfectly the period of the drafting, the presentation and the approval of the 'Integrated Energy and Climate Change Package', as presented by the European Commission on the 10th of January 2007 and as approved by the Spring European Council of the 8th and 9th of March 2007. The importance of the Commission's package and the Council's conclusions must be strongly emphasized. They set, at the EU level, legally binding targets regarding the reduction of greenhouse gas emissions, energy efficiency, renewable sources of energy, and biofuels. The Energy Action Plan 2007-2009, included in the annex to the Council's conclusions, will be the leading document for the future energy policy of the EU. The package, the conclusions and the plan, confirm once again the far-reaching ambitions of the EU in the field of climate change (policy). The most important instrument of the EU climate change policy remains the EU ETS, the European Union emission trading scheme. After the adoption of the Emission Trading Directive and its modification by the Linking Directive, the European Commission is now, on the basis of its first experiences, preparing a modification of the Emission Trading Directive. After several hearings, to be concluded by the summer of 2007, the Commission is expected to present a proposal for a directive in the autumn of 2007, aiming at, as soon as possible, approval and entry into force, and member state implementation by 2013. Next to more harmonization, simplification and enforcement, an enlargement of the Emission Trading Directive's scope of application is high on the agenda. As is well-known, air traffic is one of the strongly debated sectors to be included. The revision of the Emission Trading Directive is of great importance: it must make emission trading more effective and efficient, and more credible. After a first round of weak national allocation plans (2005-2007), the Commission is well aware of the fact that it must take a though position in the second round of national allocation plans (2008-2012), and it does, much to the dissatisfaction of some member states, which even want to challenge the allocation decision of the Commission before the European Court of Justice. We will have to wait and see. Although emission trading remains the central instrument of the EU's climate change policy, earlier this year the Commission published an interesting green paper on the use of market-based instruments in the environmental and energy policy of the EU. In the paper, an interesting link is made between energy taxation and emission trading. Obviously, in a number of situations, energy taxation is a more efficient and effective instrument than emission trading, and should, indeed, again be considered as a policy instrument in the battle against climate change. The unanimity problem in the Council obviously remains, but perhaps the track of enhanced cooperation could finally and definitely be explored.

Obviously, there are not only European developments in the climate change policy. Also at the international level a great deal of action and deliberation is being undertaken, and this within different settings: UNFCC, IPCC, World Economic Forum, G8, Security Council, etc. Several important meetings have taken place and will take place in 2007. All of them aim at getting the post-Kyoto track or the post-2012 track launched. Between 2007 and 2009, a new international framework should be negotiated and approved, in order to continue international cooperation after 2012. Due to remaining differences between the EU, the US, and also big countries like India, China and Russia, as well as the large group of developing countries, the launching of the post-2012 debate remains problematic and it is far from certain that a new and strong legal framework will be developed in time. The involvement of all countries, but based on differentiated responsibilities and capabilities, should however and anyway remain the leading principle in this debate.

Next to the European and international level, a great deal is of course happening at country level. Almost on a daily basis, individual countries announce specific targets or actions in the field of climate change policy. Reference can be made, for example, to important initiatives at state level in the USA and Australia. It is clear that in many countries climate change has taken a high position on the political agenda, and forces politicians to take action in order to respond to societal demands and concerns.

I would like to conclude with a note of caution: in several countries and organizations, we see a certain climate change hype, created and increased by movies and reports, correctly or incorrectly. As always, one must be careful with hypes, also with the climate change hype. It is clear that there is a problem and that action is needed. However, one must avoid over-reactions, certainly when it is emotionally-driven and done in a sphere of over-bidding between countries or international organizations.

Leuven, May 2007   Prof. Dr. Kurt DEKETELAERE*

---

* Full Professor of Environmental and Energy Law, University of Leuven; Visiting Professor of Climate Change Law (2007), National University of Singapore; Honorary Professor of Climate Change Law (2007-2008), University of Dundee; Senior Legal Advisor to the Flemish Minister for Energy and Environment.

ACKNOWLEDGEMENTS

The publication of this book would not have been possible without the contribution of a broad range of persons, companies and institutions.

First and foremost, we would like to thank the individual authors for producing their chapters after having presented their initial papers at the conference 'Tackling Climate Change – An Appraisal of the Kyoto Protocol and Options for the Future' of 30-31 March 2007 at the T.M.C. Asser Institute, and in some cases at the earlier conference in Siena on 'The Climate Change Regime: Outstanding Issues' held on 10-11 June 2006.

Where the Asser Conference is concerned, we would also like to express our gratitude to Lyske de Vries, Anna de la Parra, Ruben Vermeeren, Laurien de Vos and Sara Woods. All of them have been of great help throughout the organization of the conference. The Asser Conference was further realized thanks to the generous support of the Dutch Ministry of Housing, Spatial Planning and the Environment (VROM), Shell, De Brauw Blackstone Westbroek, Huglo Lepage and the media-partnerships with PointCarbon and ENDS Europe Daily/ENDS Europe Report. We owe the Dutch Ministry of Housing, Spatial Planning and the Environment our special gratitude for also substantially contributing to the publishing of this book. As far as the Sienna Conference is concerned, a special thanks goes to Cristina Carrino, Maria Chiara Alberton, Flavia Di Pilla and the Italian Ministry for the Environment, Land and Sea.

Then there is Peter Morris, who proved to be indispensable as regards the language editing of this book, and Mieke Eijdenberg, who was responsible for its subediting. We would like to thank both of them for helping to swiftly but accurately turn the contributions into a ready manuscript. Finally, Asser Press was instrumental in a rapid production process, in order for this book to be ready promptly after the last conference and in time to contribute to the ongoing debate on the post-2012 regime.

The Hague/Siena, June 2007

Wybe Th. DOUMA
Leonardo MASSAI
Massimiliano MONTINI

TABLE OF CONTENTS

Foreword   V
Kurt Deketelaere

Acknowledgements   VII

List of Abbreviations   XV

Introduction   1
Wybe Th. Douma, Leonardo Massai and Massimiliano Montini

The International Climate Regime: where do we stand?   5
Feng Gao

Part i: The implementation of the Kyoto Protocol flexible
mechanisms   11

Legal challenges in European climate policy   13
Leonardo Massai
1.   Introduction   13
2.   The European Climate Change Programme   14
3.   European Allowance Trading Directive   18
4.   The EU position on the future climate policy   23
5.   Conclusions   25

The EU Emissions Trading Directive: time for revision?   29
Nick Farnsworth
1.   The future of emissions trading: additional sectors and gases   29
1.1   The scope of the directive   30
1.2   Further harmonization and increased predictability   33
1.3   Robust compliance and enforcement   35
1.4   Links to third countries   36
2.   Conclusion   37

The role of joint implementation within the context of EU policies 39
Wytze van der Gaast
1.   Introduction   39
2.   State of play concerning JI   40
3.   Compatibility of JI with EU policies   44
3.1   JI and the *Acquis Communautaire*   44
3.2   JI and the EU Emissions Trading Scheme   47

4.   Enhancing the scope for JI through standardization of accounting
      procedures   51
4.1   Introduction   51
4.2   Grid-connected Renewable Energy Sources projects   53
4.3   Transport   55
4.4   Waste management   55
5.   Conclusion   56
References   57

The market potential of large-scale non- CO₂ CDM projects   59
Axel MICHAELOWA, Jorund BUEN, Arne EIK and Elisabeth LOKSHALL
Abstract   59
1.   Introduction   59
2.   HFC-23 destruction   61
3.   Destruction of N₂O from adipic acid production   63
4.   Destruction of N₂O from nitric acid production   64
5.   Landfill gas capture and destruction   65
6.   Coalmine and coal-bed methane capture/destruction   66
7.   Gas flaring reduction from oil production   66
8.   PFC emissions reduction from aluminium production   67
9.   Overall CER potential until 2008 and the impact on the CDM market 67
References   68

Legal Nature of Kyoto Units   71
Matthieu WEMAERE

Marketing CERs: Legal and Contractual Issues for Sellers   79
Charlotte STRECK
1.   Introduction   79
2.   CDM marketing strategy: main attention points   80
2.1   Definition of a sales strategy   80
2.2   Timing of the sale: spot and forward markets   83
3.   Existing CDM carbon contract models   85
3.1   The World Bank ERPA   86
3.2   The IETA ERPA   86
4.   CERSPA: A new open-source CDM contract   87
4.1   Contractual cornerstones   88
4.2   Events of default and remedies   91
5.   Conclusion   92

PART 2: EXPERIENCES AND PERSPECTIVES IN THE INTERNATIONAL
CLIMATE REGIME   93

The compliance regime of the Kyoto Protocol   95
Massimiliano MONTINI
1.   Introduction: dispute settlement and dispute avoidance under MEAs 95
2.   The compliance regime of the Kyoto Protocol: some basic legal issues 99
3.   The Compliance Committee and the consequences for non-compliance 103
4.   The responsibility for non-compliance under International law and EC
      law compared   106

Implementation of the Kyoto Protocol in Germany:
Designing an integrated management scheme for Greenhouse Gases   111
Michael MEHLING
1.   Introduction   111
2.   Climate policy in Germany: A model of diversity   113
2.1   Ambitious objectives – A myriad of solutions   113
2.2   A wealth of regulation – A dearth of achievement?   116
3.   Synergy or conflict? Dissecting the 'instrument mix'   118
3.1   Global Warming – A new kid on the block?   118
3.2   Internal and external conflicts – An analytical framework   120
4.   Coherence by design: Envisioning a management regime   125
4.1   The legal context – Identifying a mandate   126
4.2   Integrated Greenhouse Gas management – Clinching the objective 129
4.3   Towards a Greenhouse Gas Management Act – Specifying Design
      Elements   130
5.   Conclusion and outlook   132

The Russian Federation and the Kyoto Protocol   135
Wybe DOUMA and Daria RATSIBORINSKAYA
1.   Introduction   135
2.   From signing to ratification   135
3.   Russia's obligations and prospects under the KP   138
4.   The adoption of general implementation measures   140
5.   The Joint Implementation regime   142
6.   Concluding remarks   144

Legal and institutional barriers to Kyoto Protocol implementation
in non-Annex I countries in South-Eastern Europe and CIS   147
Marina OLSHANSKAYA
Abstract   147
1.   GHG reduction potential and the status of the CDM pipeline   147
1.1   Countries in a 'pre-market' phase: Central Asia and South-Eastern
      Europe   148

1.2   Countries in an initial stage of engagement in the carbon market:
        Southern Caucasus and Moldova   149
2.   Status of the National Institutional Framework in Eastern Europe and
        the CIS   151
3.   Trends in institutional development   154
4.   National CDM project review and approval procedures   156
5.   Conclusions   157
References   159

Developing Countries and the Post-Kyoto Regime:
Breaking the tragic lock-in of waiting for each other's strategy   161
Joyeeta GUPTA
1.   Introduction   161
2.   Challenges facing developing countries   162
2.1   Introduction   162
2.2   The negotiation challenge   162
2.3   The climate challenge   164
2.4   Inferences   167
3.   The case of China and India: Keeping the doors for entry open   167
3.1   Introduction   167
3.2   Mainstreaming in the UN   168
3.3   Encouraging local government policy   169
3.4   Litigation   171
4.   Vulnerability of developing countries (2000)   173
4.1   Introduction   173
4.2   Aid   173
4.3   Mechanisms to promote technology transfer   175
5.   Conclusions   176

Climate change, security and forests   181
Wouter VEENING

EU ETS in the post-2012 regime: lessons learned   185
Chris DEKKERS and Machtelt OUDENES
1.   Introduction   185
2.   The critical issues of emissions trading   187
3.   Factors that determine the success of EU ETS   188
4.   Lessons learned from the first trading period   189
4.1   Monitoring and reporting   190
4.2   Verification and accreditation   192
4.3   Inspection and enforcement   193
4.4   Sanctions   194
4.5   Exchange of best practice   194
5.   Other legal and technical issues of emission trading   195
5.1   Allocation   195

5.2   Administrative burden of installations   197
5.3   Legal status of an emission allowance   197
5.4   Interrelationship with IPPC directive   197
5.5   Transparency of monitored data   198
5.6   Lessons learned on consultation and cooperation   198
6.   Conclusions   199

**The São Paulo Proposal for an Agreement on Future International
Climate Policy**   201
Erik HAITES
Abstract   201
1.   Introduction   202
1.1   Medium and long-term goals   203
1.2   Commitments of Annex I/B Parties   203
1.3   Automatic extension of Annex I/B commitments   205
1.4   Economic hardship   207
1.5   Non-Annex I Parties' quantified sustainable development actions and
       'no lose' commitments   207
1.6   Graduation by Non-Annex I Parties   210
1.7   Clean Development Mechanism   212
1.8   Joint Implementation   214
1.9   Emissions trading   214
1.10   Compliance   215
1.11   Enhanced implementation of adaptation   215
1.12   Technology transfer   218
1.13   Technology research and development   218
1.14   Memoranda of Understanding extending the scope of the agreement
       with non-Parties   219
1.15   Memoranda of Understanding extending the scope of the agreement
       to special sectors and sources   220
1.16   Trade restrictions   220
1.17   Review   221
1.18   Legal form of the post-2012 agreement   221
References   222

**Stakeholder views on approaches and instruments for continued
and future climate change mitigation efforts post-2012**   223
Robert TIPPMANN
1.   Introduction   223
2.   Proposals on commitments and approaches for post-2012   223
3.   Continuation of project-based or flexible mechanisms such as the
       CDM   226
4.   Alternative, non-emission reduction target based approaches   228
5.   Negotiation arena   230

6.	Linking of different emission trading schemes	232
7.	Guiding principles of a future climate regime	234
8.	A gap after post-2012?	236
9.	Conclusions	237
References		239

The Kyoto Protocol as a pioneer among the multilateral environmental agreements 241
Michael BOTHE

1.	The problem and its remedies	241
2.	The regulatory challenges	242
2.1	The comprehensive scope of measures	242
2.2	Managing uncertainty	243
2.3	Managing complexity	243
3.	The regulatory instruments	243
3.1	Regulatory instruments in a multilevel system	243
3.2	Flexible mechanisms and 'economic instruments'	244
4.	Ensuring compliance – international administration	245
5.	Climate change under double jeopardy: technicity and universality	246

LIST OF ABBREVIATIONS

A
AAU   Assigned Amount Units
ACE   Adaption Committee of Experts
AIE   Accredited Independent Agency
AIJ   Activities Implemented Jointly
AWG   Ad hoc Working Group

B
BASA   Bilateral Air Service Agreements
BAT   Best Available Technology
BSA   Burden Sharing Agreement

C
CAA   Clean Air Act
CCS   Carbon Capture and Storage
CDM   Clean Development Mechanism
CDU   Christian Democratic Union
CENE   Centre for Energy Efficiency
CEPS   Centre for European Policy Studies
CER   Certified Emission Reduction
CERSPA   Emission Reduction Sale and Purchase Agreement
CERUPT   Certified Emission Reduction Unit Purchase Tender
CFI   Court of First Instance
CGE   Consultative Group of Experts
CIS   Commonwealth of Independent States
CKPP   Central Kalimantan Peatland Project
CMM   Coalmine Methane
COP   Conference of the Parties
COP/MOP   Conference of the Parties serving as the Meeting of the Parties

D
DANCEE   Danish Cooperation for Environment in Eastern Europe
DEPA   Danish Environmental Protection Agency
DG   Directorate General
DNA   Designated National Authorities
DSB   Dispute Settlement Body

E
EA   Environment Agency
EATD   European Allowance Trading Directive
EB   Executive Board
EC   European Community
ECCP   European Climate Change Programme
ECHR   European Convention on Human Rights
EIT   Economies in Transition
EPA   Environmental Protection Agency
ERPA   Emission Reduction Purchase Agreement

ERT   Expert Review Team
ERU   Emission Reduction Unit
ERUPT   Emission Reduction Unit Procurement Tender
ET   Emissions Trading
ETS   Emissions Trading Scheme
EU   European Union
EUA   European Union Allowances
EU ETS   European Union Emissions Trading Scheme

G
GATS   General Agreement on Trade in Services
GATT   General Agreement on Tariffs and Trade
GDP   Gross Domestic Product
GDR   German Democratic Republic
GEF   Global Environment Facility
GHG   Greenhouse Gases
GIS   Green Investment Scheme
GWP   Global Warming Potential

I
IACC   Interagency Committee on Climate Change
ICAO   International Civil Aviation Organisation
ICEP   Integrated Climate and Energy Policy
ICER   Industry Council for Electronic Equipment Recycling
ICJ   International Court of Justice
ICLEI   International Council of Local Environmental Initiatives
ICSU   International Council of Scientific Unions
IEA   International Energy Agency
IETA   International Emission Trading Association
IFDC   International Center for Soil Fertility and Agricultural Environment
IPCC   Intergovernmental Panel on Climate Change
IR   Initial Report
IRR   Internal Rate of Return
ISDA   International Swaps and Derivatives Association
IT   Information Technology

J
JI   Joint Implementation
JISC   Joint Implementation Supervisory Committee

K
KP   Kyoto Protocol

L
LCER   Long-Term Certified Emission Reduction
LDC   Least Developed Countries
LEG   LDC Expert Group
LFG   Landfill Gas
LULUCF   Land Use, Land Use Change and Forestry

M
MDG   Millennium Development Goals
MEA   Multilateral Environmental Agreements
MEDT   Ministry of Economic Development and Trade
MoEPP   Ministry of Environment and Physical Planning
MoEW   Ministry of Environment and Water
MoU   Memorandum of Understanding
MRG   Monitoring and Reporting Guidelines

N
NAP   National Allocation Plan
NAPA   National Adaptation Programmes of Action
NC   National Communication
NGO   Non-Governmental Organization

O
ODA   Official Development Assistance
OECD   Organisation for Economic Co-operation and Development

P
PDD   Project Design Document
PIN   Project Idea Note

Q
QELRC   Quantified Emission Limitation and Reduction Commitment

R
R&D   Research and Development
RGGI   Regional Greenhouse Gas Initiative
RMU   Removal Units

S
SBI   Subsidiary Body for Implementation
SBSTA   Subsidiary Body for Scientific and Technological Advice
SCM   Subsidies and Countervailing Measures

T
TBT   Technical Barriers to Trade
tCER   Temporary Certified Emission Reduction
TEAB   Technology and Environment Branch

U
UN   United Nations
UNCTAD   United Nations Conference on Trade and Development
UNDP   United Nations Development Programme
UNEP   United Nations Environment Programme
UNFCCC   United Nations Framework Convention on Climate Change
USEPA   United States Environmental Protection Agency

V
VER   Verified Emission Reduction

W
WHO   World Health Organization
WMO   World Meteorological Organization
WTO   World Trade Organization
WWF   World Wide Fund for Nature

INTRODUCTION

This publication is the result of two major events which brought together a selected group of experts in order to highlight and identify the main legal and institutional questions related to the implementation of the Kyoto Protocol. These events were the conference 'Tackling Climate Change – An Appraisal of the Kyoto Protocol and Options for the Future' held on 30-31 March 2007 at the T.M.C. Asser Institute in The Hague and the conference 'The Climate Change Regime: Outstanding Issues' organized by the University of Siena on 10-11 June 2006.

The rules and details of the international climate regime, defined after so many years of negotiations, from the UNFCCC (1992) up to the Seventh Conference of the Parties to the UNFCCC (COP7) and the first Conference of the Parties serving as the meeting of the Parties to the Kyoto Protocol (COP/MOP1) held respectively in 2001 (Marrakech) and in 2005 (Montreal), are now in place at least until the end of the first commitment period (2008-2012). Not irrelevant are the compromises made by the international community to have the system, although very often too complicated, tedious and not transparent, running and active. On this matter, it is important to remember that the current structure of the Kyoto Protocol agreed in 1997 is mainly the result of a major compromise with the group of negotiating countries led by the United States that successfully included in the final text, among other things, a five-year budgeting period instead of a one-year target or the intro-duction of flexible mechanisms instead of exclusively domestic measures. Similar concessions were the result of the Marrakech Accords in 2001: a significant in-crease in the amount of sinks allowance for the Russian Federation or the failure to agree on the binding nature of the compliance regime for Japan. The focus of the international negotiations is now on the decisions over the post-2012 phase.

The major results of the flexibility of the Kyoto Protocol are the mechanisms established within its framework, namely Emissions Trading (ET), Joint Imple-mentation (JI) and the Clean Development Mechanism (CDM). The launch of the International Transaction Log by the UNFCCC in 2006 allows for the interconnec-tion of all these mechanisms and makes this system very likely to be mimicked in the future regime, whatever this will be. Furthermore, the election of the managing bodies and the adoption of the procedures for the implementation of JI projects, as well as the first years of implementation of the European Union Emissions Trading Scheme and the Clean Development Mechanism, demonstrate the importance and the value of this new cost-effective approach towards combating climate change.

Both the above-mentioned conferences were held in the European Union. Never-theless the focus was not only on the EU but also on the position of developing states and initiatives from other parts of the world, as will be explained below. Meanwhile, the role played by the EU and its member states in the development of

W.Th. Douma et al., eds., The Kyoto Protocol and Beyond
© 2007, T·M·C·ASSER PRESS, *The Hague, The Netherlands and the Authors*

advanced legislation aimed at tackling global warming has been and will continue
to be of the utmost importance. With the introduction of the EU Emissions Trading
System linked with the international flexible mechanisms – Joint Implementation
and Clean Development Mechanisms – the EU is at the centre of this innovative
legislative and institutional framework in place for the reduction of greenhouse gas
emissions. This is confirmed by the latest data on the functioning of the so-called
carbon market: a tool where the economic and environmental aspects related to the
fight against global warming meet through the exchange of the different Kyoto
carbon reduction units. In its annual carbon market report (2007), the World Bank
estimates that the demand for JI and CDM carbon reduction units come primarily
(75%) from the EU primary sectors – those covered by the EU ETS, namely the
energy and other main industrial sectors – and from those EU member states not yet
on track with their Kyoto Protocol targets. Not only is the EU at the forefront on the
demand side and it is the main actor in the field of international cooperation for the
reduction of the level of greenhouse gas emissions. On the supply side, together
with Russia and Ukraine, the former Central and Eastern European countries –
nowadays EU member states like Bulgaria, Romania, the Czech Republic, Poland,
Hungary and others – are ready to offer the required assistance to the demand side,
both in terms of cheaper conditions for projects aimed at the reduction of greenhouse gas emissions and in terms of the availability of Kyoto units to trade. Finally,
strictly looking within the EU, the decisions over climate and renewable energy
policy included in the EU presidency conclusions adopted on 8-9 March 2007 demonstrate the level of commitment of the European institutions and governments
towards climate change. It will be the responsibility of the member states to implement these decisions correctly by the adoption of adequate policies and measures at
the national level.

The publication includes the main presentations and discussions held in the two
mentioned conferences, where academics, policy makers, consultants and different
stakeholders met to discuss the current and future design of the international climate regime, especially from a legal and regulatory perspective. Although not all
contributions from the invited speakers are incorporated, the editors were able to
provide a comprehensive volume reflecting the main ideas and outcomes of the two
meetings.

The volume starts with the contribution by Gao, who gives a very comprehensive and concise insight into the current status of the international climate regime,
namely the level of greenhouse gas emissions, the reporting requirements, the compliance system and the implementation of the flexible mechanisms.

The first block of papers included focus in particular on the main legal and regulatory aspects related to the implementation of the flexible mechanisms. It starts
with the assessment of European Climate Policy, in particular with the main legal
considerations concerning the recent developments in the EU Emissions Trading
Scheme, the European Climate Change Programme, and the future of EU climate
policy (Massai) and the revision of the EU Emissions Trading Directive (Farnsworth).

The state of play in the field of JI is provided by Van der Gaast, who first identifies the recent developments in the infrastructure and rules for JI at the UNFCCC

level, and then considers the relation between the implementation of JI and the EU rules. The focus on CDM is based on the contribution by Michaelowa which gives a comprehensive overview of the latest data on the actors and the types of projects concerned.

The first section of the book ends with two papers dealing with specific legal aspects of climate change regimes, namely by Wemäere dealing with the legal nature of Kyoto carbon reduction units and by Streck discussing the development of a new carbon contract template.

The second part of the volume considers the international climate regime both looking at experiences in the implementation of the Kyoto Protocol's ideas in different countries and perspectives in respect of the post-2012 phase.

Montini starts with an evaluation of the Kyoto Protocol compliance regime, in particular highlighting the peculiarities of this system in comparison with other multilateral environmental agreements, the binding force of the regime and the future steps to be taken.

The case studies on the implementation of the Kyoto Protocol at the country level start with Mehling who introduces the main features of a management system for the atmosphere in Germany. He discusses the incoherencies incumbent in this system, and its political and legal constitutional aspects, before setting out to sketch the need for a more comprehensive management scheme for greenhouse gas where existing and new instruments would be brought together under an umbrella law, tentatively designated as the 'Greenhouse Gas Management Act'. Mehling relates the content of his ideas to earlier experience with environmental codification in Germany. From the side of the editors, we merely hope that the adoption of such a law would not entail the same amount of time as has been involved in realizing a German General Environmental Law (*Bundesumweltgesetzbuch*), which has been under discussion for over 17 years and is still to be adopted.

Douma and Ratsiborinskaya present the case of the Russian Federation, in particular focusing on the struggle to get this country to ratify the Kyoto Protocol and what could be learned from this when it comes to the negotiations on the post 2012-regime. They also discuss the first pieces of legislation with regard to Joint Implementation in Russia that were agreed upon at the end of May 2007. Where Mehling finds that in Germany legislation might often have been adopted too hastily, resulting in a lack of cohesion and systematization in German climate policy, Douma and Ratsiborinskaya put forward that the relatively late adoption of climate change legislation in Russia does not necessarily warrant better results.

The legal and institutional aspects of the Kyoto Protocol's implementation in the Western Balkan and CIS countries are investigated by Olshanskaya who considers, in particular, the institutional constraints required for the implementation of CDM and JI, as well as the rules for the approval of those types of projects in the region.

The section on the post-2012 phase includes the contribution by Gupta who considers the position of developing countries in the post-2012 regime. She focuses on the possibility of China and India accepting some form of quantitative commitments under the evolving climate change negotiations; and on the impact of climate change and climate change politics on the larger group of developing countries. Veening tackles the issue of climate change and sustainable development, in connection with security questions and forestry / peat management needs, providing a valuable insight into the highly significant steps that should be taken to combat climate change effectively.

Dekkers and Oudenes provide an insight into the lessons learned in the first trading period of the EU Emissions Trading System and how these should be reflected in the post-2012 discussions, especially in respect of monitoring and reporting, verification and accreditation, enforcement and inspection. Haites describes the Sao Paulo proposal aiming at the creation of a coherent and balanced package of ideas for the discussions on the post-2012 regime. The section closes with Tippmann providing an analysis on a series of stakeholders' views on approaches and instruments for continued and future climate change mitigation efforts post-2012. The book ends where the Asser Conference began, namely with the contribution of Michael Bothe, who considered the 'pioneer' character of the Kyoto Protocol among the multilateral environmental agreements

Of course, the two conferences could not deal with all the issues surrounding climate changes. We are confident, however, that the various contributions in this publication drawn from these conferences can contribute to the debate on the formidable challenges that lie ahead in designing an effective and workable post-2012 regime.

The Hague, June 2007   Wybe DOUMA and Leonardo MASSAI
                                        T.M.C. Asser Instituut

                                        Massimiliano MONTINI
                                        University of Sienna

THE INTERNATIONAL CLIMATE REGIME: WHERE DO WE STAND?

Feng Gao*

It is a pleasure and an honor to contribute to this publication on behalf of the secretariat of the United Nations Framework Convention on Climate Change (UNFCCC). It is indeed critical to give very serious consideration to the implementation of the Kyoto Protocol now, as we are quickly approaching the first year of the first commitment period in 2008. The theme of this book is significant because only when we make the Kyoto Protocol work, can we talk about deeper cuts in greenhouse gas (GHG) emissions after 2012, when the Kyoto Protocol needs to be extended to its next commitment period.

As requested, I will focus on 'Where Do We Stand?' with regard to the Kyoto Protocol, and also touch briefly upon 'Where Do We Go?' towards the end. Within the limited time frame I can only cover a number of key issues related to the implementation of the Protocol. To make the theme clearer, it is perhaps necessary to provide some information on the process and issues under the Convention.

The starting point should, I believe, be a brief overview of the current GHG emissions by Parties included in Annex I to the Convention (Annex I Parties): in 2006, the report by the UNFCCC secretariat on the inventories of Annex I Parties for the period 1990-2004 shows that total aggregate GHG emissions, including emissions from land use, land-use change and forestry (LULUCF) for Annex I Parties as a whole, had decreased by 4.9 per cent between 1990 and 2004. For Annex I Parties with economies in transition (EITs) GHG emissions with LULUCF decreased by 44.8%. For the Annex I Parties that are not EITs, GHG emissions with LULUCF increased by 12.1%. The Subsidiary Body for Implementation (SBI) noted in one of its conclusions that the most recent GHG emission projections suggest that between 2000 and 2010 the total GHG emissions of Annex I Parties could increase by about 13%, despite the measures currently in place. The SBI urged Annex I Parties to make additional efforts to reverse this trend.

As we are aware, the implementation of the commitments of Annex I Parties that are Parties to the Kyoto Protocol is first based on a set of requirements of reporting from those Parties. In this respect, the Protocol builds on the requirements of the Convention. It adds supplementary requirements to the two regular, ongoing reporting requirements for Annex I Parties: the annual report and national communi-

---

* Legal Affairs, UNFCCC.

W.Th. Douma et al., eds., The Kyoto Protocol and Beyond
© 2007, T·M·C·ASSER PRESS, *The Hague, The Netherlands and the Authors*

cations (NCs). The fourth national communications (NC4s) were due on 1 January 2006. In addition, these Parties were to submit an initial report (IR) on 1 January 2007. In 2007, a review of the IR is particularly important for the calculation of an Annex I Party's assigned amount and for establishing mechanisms eligibility before 2008.

The performance of most Annex I Parties in this respect has been fairly good. However, areas of concern and possible future questions of implementation are emerging, such as non-submission and late submission: to date, two Parties have still not submitted NC4s, and another two have failed to submit IRs on time. Sixteen reports from the expert review teams (ERTs) have so far been received by the Compliance Committee, and all are on NC4s. No IR review reports have been received. Issues have also arisen with respect to incomplete inventories, as well as other issues related to specific methodologies and good practice guidelines. However, to date, those Parties appear to have been able to provide sufficient additional information and clarifications to satisfy ERTs. So far, no questions relating to implementation or adjustment recommendations have been contained in any ERT report submitted to the Compliance Committee.

The compliance system is one of the crucial components to promote compliance with the commitments under the Kyoto Protocol for the Annex I Parties that are Parties to the Protocol. By its decision 27/CMP.1, the Conference of the Parties serving as the meeting of the Parties to the Kyoto Protocol at its first session (CMP 1) in 2005 adopted the compliance procedures and mechanisms under the Kyoto Protocol, subject to a decision of the CMP on the issue of an amendment to Article 18 of the Kyoto Protocol. In 2006, CMP 2 adopted the Committee's rules of procedure, making it ready to fully function. In the same year, the Committee, upon a request from South Africa on behalf of the Group of 77 and China, considered the issue of the non-submission and late submission of the reports demonstrating progress under Article 3.2 of the Kyoto Protocol from some of the Annex I Parties that are Parties to the Kyoto Protocol, and was unable to decide whether or not to proceed with the question of implementation. This has been reflected in the first report of the Committee to CMP 2. A decision by the SBI on this issue is still pending.

For Annex I Parties that are Parties to the Kyoto Protocol to implement their Kyoto commitments at the lowest possible cost, the three flexibility mechanisms have proven to be powerful policy tools. The clean development mechanism (CDM) has already shown its strength in not only facilitating compliance but also in financing projects reducing GHG emissions in developing countries. With 561 projects registered and more than twice as many (1,600) in the pipeline, the CDM is expected to deliver emissions reductions of more than 1.9 billion tonnes of CO_2 equivalent. The CDM has been recognized as the most effective mechanism in promoting the widest possible mitigation actions in developing countries in the present and for the future. In the meantime, ideas on issues such as a possible simplification of procedures, streamlining some key concepts, and a reduction in transaction costs, have been widely explored.

On the other hand, joint implementation (JI) has moved quickly since its formal launch at CMP 1 from developing the process to handling projects, and it opened its verification process in October 2006. To date, 44 JI project development documents from nine host countries have been submitted for public comment. They would result in a reduction of 88 million tonnes of CO_2 equivalent over the commitment period. So far the Russian Federation has emerged as the dominant host Party for JI projects (23). This is only the start – a great potential is there. Overall, JI could generate several hundred million tonnes of CO_2 reductions by the end of the first commitment period. However, the magnitude of JI will depend on how quickly host and partner countries move to develop projects, with 2007 clearly representing an important window of opportunity.

The Kyoto Protocol may also show its potential in assisting developing countries in their adaptation efforts through the Adaptation Fund, which is funded by a 2% share of the proceeds of the issued certified emission reduction units. The CMP, at its second session, requested the SBI to develop recommendations to CMP 3 with the aim of adopting a decision on the following issues: (a) eligibility criteria; (b) priority areas; (c) monetizing the share of proceeds; and (d) institutional arrangements.

The main contentious issue remains the institutional options. Some developing countries propose institutions other than the Global Environment Facility to host the fund. The discussion is linked to how some developing countries see the operation of financial cooperation beyond 2012. More and more proposals refer to ideas on how to create sources of funding that are not dependent on voluntary contributions from donor countries only, but rather come from an expansion of the share of proceeds concept to other mechanisms (JI and emissions trading), creating additional carbon taxes (e.g., aviation taxes) or exploring 'old ideas' such as the 'compliance fund' from the Brazilian proposal.

The implementation of the Kyoto Protocol has a direct bearing on its future. It is obvious that despite the ambitious and sophisticated policies and measures, commitments and institutions that governments have put in place, significant additional action is needed to reduce emissions after 2012 and to support adaptation to climate change. The effects of the Kyoto Protocol are limited, as Australia and the United States of America are not Parties to the Protocol and developing countries do not undertake quantitative emission reduction and limitation targets; hence the Protocol is able to address only 30% of global emissions.

To address this problem, a two-track process was launched at CMP 1 in December 2005: the dialogue on long-term cooperative action to address climate change by enhancing the implementation of the Convention (the Dialogue), and negotiations on new commitments after 2012 for industrialized countries to implement Article 3.9 of the Kyoto Protocol. Decision 1/CMP.1 established the Ad Hoc Working Group on Further Commitments for Annex I Parties under the Kyoto Protocol (AWG) for this purpose. The COP agreed that this group should aim to complete its work and have its results adopted as early as possible and in time to ensure that there is no gap between the first and second commitment periods of the Protocol.

The AWG is open-ended and is to report each year to the CMP on the status of its work.

The AWG agreed to focus its work for 2007 on the first task of its workplan covering the analysis of mitigation potentials and ranges of emission reduction objectives of Annex I Parties that are Parties to the Kyoto Protocol. At its third session, the AWG will, through a round table, address a broad range of aspects of the mitigation potential of policies, measures and technologies, including some showcase examples. The AWG plenary would select some of the analytical elements that lend themselves to quantifying possible ranges of emission reductions for Annex I Parties that are Parties to the Kyoto Protocol, without going into specific national figures.

At its fourth session in August 2007, the AWG should aim to finalize the discussions on the mitigation potentials and the identification of ranges of emission reductions and will revisit the workplan with the aim being to agree on the tasks for 2008 at the resumed AWG 4 in Bali in December.

On the other hand, the Dialogue is to focus on four thematic areas, namely: advancing development goals in a sustainable manner; addressing action on adaptation; realizing the full potential of technology; and realizing the full potential of market-based opportunities.

Two workshops under the Dialogue have already taken place and in 2007 the Dialogue will be completed. The third workshop will address the topics of adaptation and technology. The fourth and final workshop will address overarching and cross-cutting issues, including finance, and proposals for further actions, activities and approaches. The co-facilitators are to finalize their written report, which will be presented at COP 13. This report is likely to attempt to provide elements for the future consideration of cooperative action on climate change and possible linkages between these elements, drawing upon the diversity of views presented at all four workshops.

Because the Dialogue ends in 2007, a new round of negotiations on climate change policy beyond 2012, which is to be much broader than just the extension of the Protocol to its second commitment period, has to be initiated by the end of 2007. It will require at least two years to reach an agreement, which could then be ready by 2009 or 2010. There seems to be a very narrow window for urgent actions to speed up negotiations for the post-2012 agreement to be submitted to national governments for ratification if a gap between the first commitment period and the next commitment period of the Protocol is to be avoided.

In this respect, an important factor needs to be considered, that is, the review of the Kyoto Protocol. Article 9 of the Protocol requests the CMP to periodically review the Protocol in the light of the best available scientific information and assessments on climate change and its impacts, as well as relevant technical, social and economic information. It requests the first review to be carried out at CMP 2. Further reviews are to take place at regular intervals.

In accordance with Article 9, the first review of the Kyoto Protocol took place at CMP 2. It concluded that the Protocol has provided developed Parties with an opportunity to take the lead in combating climate change and that it has fostered cooperative action. It acknowledges that a number of elements could be enhanced, in particular adaptation. The second review of the Protocol will take place at CMP 4 (2008).

The second review of the Kyoto Protocol has a strong link to the work of the AWG, as both relate to the Protocol architecture. While the AWG has a relatively narrow mandate, Article 9 provides a broader scope to amend other aspects of the Protocol. However, the first review of the Protocol did not provide guidance on the 'appropriate action' required under Article 9.1 of the Protocol, and, thus, not much can be concluded at this stage. The review also provides for a bridge to the Convention process, as it calls for coordination with pertinent reviews under the Convention.

The foregoing has outlined the main aspects of 'Where Do We Stand?', and a number of the elements are actually key steps determining 'Where Do We Go?'. It is not difficult to see that an amendment to the Kyoto Protocol will clearly be part of the overall package of the future global climate regime. This regime, as pointed out on several occasions by the Executive Secretary of the UNFCCC secretariat, Mr. Yvo de Boer, will need to aim at a broadened effort to allow all aspects of a global solution to climate change to be addressed, including:

- A long-term global response to climate change in line with the latest scientific findings and compatible with long-term investment planning needs of business;
- Deep emission cuts by industrialized countries, which must continue to take the lead in line with their historic responsibility and economic capabilities;
- Further engagement of developing countries, in particular those whose emissions will soon significantly contribute to atmospheric concentrations;
- Incentives for developing countries to limit their emissions while safeguarding their goals of economic growth and poverty eradication;
- Flexibility through an enhanced carbon market to ensure the most cost-effective mitigation activities and to mobilize financial resources and incentives to developing countries.

Within the above framework, this regime would also need to address a set of key issues in order to strengthen or promote global cooperative actions against climate change and to achieve sustainable development in all countries, in particular developing countries. These key issues include:

- The urgent need to advance adaptation to climate change, particularly in countries which are most vulnerable to the impacts of climate change, such as the least developed countries and small island developing States;
- An innovative approach to enhance international technology cooperation between industrialized and developing countries that would constitute a major push to make existing climate-friendly technologies economically viable and deployed rapidly, and to generate new technologies;

- Widened and deepened information-sharing and data collection, as well as innovative methodologies and guidelines to ensure that necessary financial and technical support is made available to developing countries;
- Policy directions for deforestation in many developing countries;
- Ways to allow the proactive involvement of the private sector, and to support
  public-private partnerships in addressing climate change.

These elements, among others, will come together to build up a possible global
regime on climate change beyond 2012. There are a number of scenarios for the
final agreement on these elements, such as a minimum amendment or a maximum
amendment to the Kyoto Protocol, or an entirely new legal instrument under the
Convention, in parallel with the Protocol.

Whatever the result, the leadership provided by the Annex I Parties, including
the faithful implementation of the Kyoto commitments by those Parties to the Kyoto
Protocol among them, will no doubt be at the heart of, or even be a prerequisite to,
the success of the global regime against climate change. It is in this context that I
would like to congratulate the Asser Instituut for its significant contribution to the
serious consideration of the implementation of the Kyoto Protocol, the very first
step in our common cause of mankind against climate change.

Part 1

THE IMPLEMENTATION OF THE
KYOTO PROTOCOL FLEXIBLE MECHANISMS

LEGAL CHALLENGES IN EUROPEAN CLIMATE POLICY

Leonardo Massai*

1. INTRODUCTION

The European Community (EC), the legal entity representing the EU in the international climate regime, ratified the Kyoto Protocol on 31 May 2002 with Council Decision 2002/358/EC which certifies the Community global quantified GHG emissions reductions of 8% under the KP.[1] The EC and its member states are jointly committed to the UNFCCC and to the Kyoto Protocol, respectively, according to Article 4(6) and 24(2), and Article 4. The international treaty regime constitutes a mixed agreement according to which the EC and its member states have shared responsibility and are jointly liable. With regard to the quantified emission limitation and reduction commitments (QELRC) established by the Kyoto Protocol, the overall EC target has been redistributed among the old member states (EU 15)[2] by the Burden Sharing Agreement (BSA) – the *EU bubble* – which gives each country a differentiated reduction target.

European policy aiming at the curbing of global warming has been characterized since the beginning of the nineties by the adoption of advanced and specific secondary legislation as well as by the definition of various strategies and programmes including policies and measures for the reduction of greenhouse gas emissions. The main strategy document established at the European level is the European Climate Change Programme (ECCP) launched in 2000 with Commission Communication COM (2000) 88[3] and aimed at the reduction of CO_2 emissions, as well as the promotion of renewable energy and energy efficiency.

---

* T.M.C. Asser Institute, The Hague. The author would like to thank Wybe Douma for reviewing and commenting on this paper.

[1] Council Decision 2002/358/EC of 25 April 2002 concerning the approval, on behalf of the European Community, of the Kyoto Protocol to the United Nations Framework Convention on Climate Change and the joint fulfillment of commitments thereunder, OJ L 130/1, 15.05.2002.

[2] At the moment of writing, due to the enlargements of 1 May 2004 (Cyprus, the Czech Republic, Estonia, Hungary, Latvia, Lithuania, Malta, Poland, Slovakia and Slovenia) and of 1 January 2007 (Bulgaria and Romania), the EU is composed of 27 member states. In the course of this chapter, if not otherwise specified, new member states refer to the above-mentioned 12 countries which have recently joined the EU.

[3] Commission Communication from the Commission to the Council and the European Parliament on EU policies and measures to reduce greenhouse gas emissions, COM(2000) 88, Brussels, 8 March 2000.

W.Th. Douma et al., eds., The Kyoto Protocol and Beyond
© 2007, T·M·C·ASSER PRESS, *The Hague, The Netherlands and the Authors*

The basic goals of European climate policy are:

– The promotion of energy efficiency;
– The promotion of renewable sources;
– Cooperation with the industrial sector, NGOs and stakeholders;
– The implementation of the flexible mechanisms established by the Kyoto Proto-
  col, such as Emissions Trading, Joint Implementation and the Clean Develop-
  ment Mechanism;
– An advanced transport policy.

In early 2007 the Commission adopted the so-called 'energy package', namely the
Communication 'An Energy Policy for Europe'[4] which set the new European strategy on energy policy and, among other things, on energy efficiency, biofuels and
renewables. The main content of the energy package has been confirmed by the
European Council of 8-9 March 2007[5] which adopted the following:

– Binding target for the EU member states to reduce GHG emissions by at least
  20% by 2020 in respect of 1990 levels;
– Binding target for all industrialized countries to reduce GHG emissions by 30%
  by 2020 if other developed countries will agree;
– Overall objective for all developed countries to cut their GHG emissions by 60-
  80% by 2050;
– Binding target to increase the use of renewable energy by 20% by 2020 (this to
  be divided into binding national targets for every member state);
– Binding target for the share of biofuels in the EU transport fuel consumption of
  10% by 2020

This paper will address the latest developments and the legal challenges in European climate policy, namely the new phase of the ECCP, the implementation of the
EU Emissions Trading Scheme (EU ETS)[6] and the European position on the future
of the international climate regime.

2.   THE EUROPEAN CLIMATE CHANGE PROGRAMME (ECCP)

The main goal of the ECCP is the identification of adequate policies and measures
to be adopted at the European Community level in order to reduce the level of GHG
emissions. The ECCP work is structured on several working groups composed of

---

[4] Communication from the Commission to the European Council and the European Parliament,
An energy policy for Europe COM(2007) 1, Brussels 10 January 2007.
[5] Presidency conclusions of the Brussels European Council of 8-9 March 2007.
[6] Directive 2003/87/EC of the European Parliament and of the Council of 13 October 2003 establishing a scheme for GHG emission allowance trading within the Community and amending
Council Directive 96/61/EC, OJ L 275, 25.10.2003 pp. 32-46.

representatives of the different Directorates General of the Commission, member states, the private sector, NGOs and other stakeholders. The Second European Climate Change Programme (ECCP II) (2000-2004) has provided several inputs for the adoption of Community legislation in the field of flexible mechanisms, energy supply, energy consumption, transport, industry, research, agriculture, forest-related sinks, sinks in agricultural soils, fluorinated gases, industry and waste.

ECCP II was launched by the European Commission on 24 October 2005 with Communication COM(2005) 35, 'Winning the battle against climate change'[7]. ECCP II is expected to provide background information and policy documents for the adoption of specific legislation in the following topics:

- ECCP I review (with 5 subgroups: transport, energy supply, energy demand, non-CO2 gases, agriculture);
- Aviation;
- CO_2 emissions and cars;
- Carbon capture and storage;
- Adaptation to climate change;
- Review of the EU ETS.[8]

In particular, on carbon capture and storage (CCS) a European carbon capture development group was launched on 2 December 2005 by the Commission Directorate General on Research. This group is called the Zero Emission Technology Platform and it aims at the coordination of research and activities in the sector of carbon capture and storage at the European level. The Zero Emission Technology Platform is composed of representatives of the industrial sector, EU and government officials, academics and NGOs. On 12 September 2006 it adopted two carbon capture and storage action plans aimed at the promotion of technology in this area:

- Policy framework recommendations to accelerate the market developments in the sector;
- Research recommendations to foster research on CCS.

The European general strategy towards CCS was part of the energy package adopted by the European Commission in January 2007. In particular, Commission Communication COM(2006) 843[9] paves the way for the necessary adjustments in the member states' legal frameworks to ensure that CO_2 emissions are stored in a safe way and to overcome the existing barriers created by international law.[10]

---

[7] Communication COM(2005) 35, Winning the battle against climate change, Brussels, 9 February 2005.

[8] See the discussion on the European Allowance Trading Directive (EATD) below.

[9] Commission Communication Sustainable power generation from fossil fuels: aiming for near-zero emissions from coal after 2020 COM(2006) 843, Brussels 10 January 2007.

[10] The London Protocol on dumping of waste at sea and the OSPAR Convention (on the protection of the marine environment of the North-East Atlantic).

On aviation, considering that the release in the atmosphere of greenhouse gas emissions generated by air transport is substantially increasing,[11] the reduction of these emissions has become one of the key priorities of European climate policy. The European Commission confirmed in the Communication 'Reducing the Climate Change Impact of Aviation' COM(2005) 459[12] that the inclusion of the aviation sector in the EU ETS could represent one of the most efficient ways for the mitigation of the sector's negative impacts on climate. The Environment Council of 2 December 2005 backed the proposal of the Commission and called for the extension of the obligations under the Emissions Trading directive to intra-EU and non-intra-EU flights, as well as to the coverage of all GHG emissions generated from air transport. On 4 July 2006 the European Parliament adopted a non-binding resolution urging the establishment of a carbon trading scheme for the aviation sector separate from the EU ETS as well as the adoption of a tax on kerosene or aviation fuel. According to the Parliament, this trading system should be independent from the EU ETS, should cover all flights from and to the EU and, finally, should be based on the distribution of allowances to air companies through auctioning. The debate among European institutions on measures to combat the pollution from air transport contributed to the release of a proposal for a directive amending Directive 2003/87/EC so as to include aviation activities in the EU ETS by the Commission on 20 December 2006.[13] The proposal calls for the inclusion in the EU ETS of GHG emissions from EU intra-flights from 2011 and all flights to and from EU airports from 2012. The amendment to Directive 2003/87/EC is intended to enter into force in the trading period from 2013 onwards. Furthermore, the European Commission is also investigating the impact of GHG emissions from shipping and the potential inclusion of this sector in the EU ETS along with air transport.[14] In this case the development of new allocation methods would be required as well as the establishment of CO_2 emission caps for all ships passing through European ports.

The importance of adaptation to the adverse effects of climate change is, together with the mitigation of GHG emissions in industrialized countries, one of the two pillars of the international climate regime. With the levels of GHG emissions constantly increasing, the focus of the international climate negotiations under the Kyoto Protocol, as well as the European debate on climate policy, is shifting more and more towards the problem of adaptation. Climate change is supposed to have a negative impact not only on developing countries, but also on the EU economy, as well as on other issues such as water, agriculture, forestry, industry and biodiversity

---

[11] See Aviation and Global Atmosphere, IPCC report, Geneva 1999.

[12] Communication Reducing the Climate Change Impact of Aviation, COM(2005) 459, Brussels, 27 September 2005.

[13] Proposal for a Directive of the European Parliament and of the Council amending Directive 2003/87/EC so as to include aviation activities in the scheme for greenhouse gas emission allowance trading within the Community COM(2006) 818, Brussels, 20 December 2006.

[14] Greenhouse Gas Emissions for Shipping and Implementation Guidance for the Marine Fuel Sulphur Directive, CE Delft 2006.

to urban life. The European climate policy on adaptation is aimed at the definition of a policy strategy to adapt to the impacts of climate change and to assist local, regional and national communities. The ECCP II working group on adaptation is therefore committed to the assessment of the impacts of climate change towards the water and marine resources, coastal zones and tourism, human health, agriculture and forestry, biodiversity, regional and urban planning, public and energy infrastructure, structural funds, development cooperation, and the role of the insurance industry. Furthermore, the working group on adaptation will contribute to the development of a green paper on the EU role in climate change adaptation, to the promotion of research into adaptation and mitigation options and to the dissemination of information on this topic within Europe.

On CO_2 emissions from road vehicles, since 2004 the European institutions have engaged[15] in a constant effort to promote a shift from the approach of the late nineties based on the voluntary agreements between the European Commission and the associations of car manufacturers,[16] to a more rigid approach to tackle the problem of the reduction of CO_2 emissions from road vehicles. In particular, two measures identified by the European Commission could match this aim, namely the introduction of annual circulation taxes and the requirement that member states shall ensure that at least 50% of car taxes are based on vehicles' CO_2 emissions by 2010,[17] as well as the introduction of legally binding targets for the reduction of CO_2 emissions. In particular, in the latter case, the recent debate at the EU level over the necessity of more stringent measures to reduce the impact of road transport on climate change, have shown the difficulty in agreeing on the extent of these new obligations. On 7 February 2007, the Commission presented the new plan for a cleaner automobile industry.[18] This plan sets binding CO_2 emission limits for the average of the CO_2 emissions of new cars and vans by 2012 equal to 130 grams per kilometer or less. The outcome is a compromise between the European Commission and the automobile industry. The latest data on the contribution of road transport to global warming, as well as the plan of the Commission, do show the failure of the approach based on voluntary agreements with car producers' associations.[19]

---

[15] Communication from the Commission to the Council and the European Parliament, Implementing the Community Strategy to Reduce CO_2 Emissions from Cars: Fourth annual report on the effectiveness of the strategy (Reporting year 2002) COM(2004) 78, Brussels 11 February 2004.

[16] The EC has signed three voluntary agreements establishing quantified CO_2 emissions reduction goals for the average of new passenger cars sold in the European Union, i.e., 140 gCO_2/km (to be achieved by 2009 by JAMA and KAMA and by 2008 by ACEA).

[17] Proposal for a Council Directive on passenger car related taxes and draft European Parliament legislative resolution on the proposal for a Council directive on passenger car related taxes COM(2005) 261, Brussels, 5 July 2005.

[18] Communication from the Commission to the Council and the European Parliament, Results of the review of the Community Strategy to reduce CO_2 emissions from passenger cars and light-commercial vehicles COM(2007) 19, Brussels, 7 February 2007.

[19] In 2006 the average CO_2 emissions per kilometer were 160 grams instead of 140 as established under the mentioned voluntary agreements.

3.   EUROPEAN ALLOWANCE TRADING DIRECTIVE (EATD)

The European Allowance Trading Directive (EATD) 2003/87/EC established the world's first regional emissions trading scheme. The EATD sets up a domestic, mandatory and entity-based cap-and-trade system, compatible with the international market created under Article 17 of the Kyoto Protocol. The system relies on the distribution – for free or by auctioning – of European Union Allowances (EUAs) equal to the emissions of 1 tonne of CO_2 equivalent to the installations falling under the categories indicated in Annex I to the EATD. Installations with a CO_2 emission level below their assigned pollution amounts may sell emission rights to installations that are in danger of exceeding their quotas. Operators that do not surrender sufficient allowances in a defined period are subjected to financial penalties. The system runs for two different periods: 2005-2007 as a first learning-by-doing phase and the second phase in 2008-2012 in conjunction with the Kyoto Protocol first commitment period. Under Article 9 of the EATD, member states are required to prepare and submit National Allocation Plans to the Commission respectively by 31 March 2004 (NAP I) and by 30 June 2006 (NAPs II). NAPs have to be designed in accordance with the criteria outlined in Annex III of EATD and have to be approved by the Commission. The NAPs contain details of the member states' caps to the GHG emissions of the EU ETS sectors and therefore are important to set the EU and member states' strategy towards the Kyoto Protocol targets. Member states have encountered several delays in the submission of the first and second phase NAPs. Moreover, in many cases the Commission rejected these documents and asked the national authorities to adjust the NAPs in accordance with Annex III criteria. The Commission has started infringement procedures in several cases under Article 226 EC Treaty for the failure by the member states to implement correctly Directive 2003/87/EC. In the first phase, the Commission rejected 13 NAPs and requested a cut of in total 4% (290 MtCO2eq.) from the proposed quantity of allowances.

Although the European Commission released on 9 January 2006 a Communication including guidelines for the preparation of NAPs II[20] based on the experience of the first phase, nearly all 25 EU member states missed the deadline for the submission of the second phase NAPs[21] and 'pre-infringement letters' were sent by the Commission to 14 member states. On 12 October 2006, the European Commission, with the political support of the European Parliament, announced infringement proceedings against Austria, the Czech Republic, Denmark, Hungary, Italy, Portugal, Slovenia and Spain for their failure to submit the NAPs in time. As in the first phase, it is very unlikely that the infringement procedure will reach the final stage.

The main legal challenges encountered by the EU institutions and the member states in the first two years of the implementation of the EU ETS are.[22]

---

[20] Communication on 'Further guidance on allocation plans for the 2008 to 2012 trading period of the EU Emission Trading Scheme' COM(2005) 703.

[21] Estonia was the only member state to notify the NAP on time.

[22] EEA Technical report No. 4/2007. Application of the Emissions Trading Directive by EU member states – Reporting year 2006, European Environment Agency.

- Decentralized implementation and administration of the EU ETS in several states where regional and local authorities are involved, depending on the constitutional structure of the state. For instance, in Spain powers related to the implementation of climate instruments are divided among central, regional and local authorities; in Belgium, competence over the EU ETS is allocated to the two Flemish and Walloon regions. Decentralization sometimes contributed to a certain degree of uncertainty among the EU institutions and the member states, in particular in relation to which are the competent authorities to be addressed for the implementation of the EU ETS;
- The definition of combustion installations: several member states encountered problems in the identification of the combustion installations according to the definition of Annex I of Directive 2003/87/EC. In particular, the Commission required France to amend its NAP for a failure to include certain combustion installations in the list;[23]
- The legal nature of EUAs: no uniformity in the legal treatment of EUAs has been achieved within the member states and while a few states consider the EUAs as financial instruments, others consider them as normal commodities;
- Malta and Cyprus: the fact that these two EU member states are still non-Annex I parties of the UNFCCC raises a problem related to the adjustments of AAUs in the period 2008-2012 due to the transfer of EUAs among member states, since non-Annex I parties are not able to issue AAUs.

The role of the European Commission in ensuring the correct implementation of the EATD is essential. The problems of the Commission in this respect are of a different nature. At first, the lack of adequate structure and the need for a more decentralized administration. This is particularly true in the case of the EU ETS, where the Commission has to cope with the national, regional and local authorities in the member states in order to verify and monitor the correct implementation of the EU ETS. Furthermore, the infringement procedure designed under European Community law (Articles 226 and 228 EC Treaty) appears to be too slow (an average of 2-4 years is required for the formulation of a judgement by the ECJ) and too centralized to ensure the effective enforcement of the EU ETS.[24] Apart from the infringement procedure, the correct implementation of the EU ETS could also be ensured by the direct enforceability of EU legislation by citizens, NGOs and/or companies. Theoretically, a European company included in the EU ETS could sue a member state for a failure to implement the EATD correctly, i.e., by failing to submit the NAP on time, provided that there is a direct causal link between the

---

[23] Arts. 1 and 2 of Commission decision of 20 October 2004 concerning the national allocation plan for the allocation of GHG emission allowances notified by France in accordance with Directive 2003/87/EC of the European Parliament and of the Council, Brussels, 20 October 2004, C(2004) 3982/7.

[24] Eritja, Mar Campins, 'Reviewing the challenging task faced by Member States in implementing the Emissions Trading Directive: issues of Member States liability', in Peeters, Marjan and Deketelaere, Kurt, *EU Climate Change Policy*, Edward Elgar, Cheltenham, 2006.

conduct of the member state and the rights of individuals. In other words, the plaintiff should prove that the negligence of the national authorities caused direct damage to the company, for instance in terms of hampering full access to the EU market. In practice, it seems unlikely that such cases would be successful.

The Linking Directive 2004/101/EC[25] amends Directive 2003/87/EC and introduces the direct linking of Emission Reduction Units (ERUs) generated by JI projects and Certified Emission Reductions (CERs) from CDM credits to the EATD. Directive 2004/101/EC allows for:

- the use of CERs (CDM) in the EATD from 2005 and ERUs (JI) from 2008;
- no formal limitation of the quantity of credits to be included in the EATD but governments are required to consider the issue of supplementarity;
- specific provisions on supplementarity to be included in NAPs for the second phase (2008 to 2012);
- exclusion of credits from nuclear projects and from sinks in the first phase;
- hydro projects to be implemented following international rules on dams.

With the Linking directive, the EU ETS becomes the first regional carbon market where all the Kyoto units, namely AAUs, Removal Units RMUs, CERs, ERU and temporary CERs (tCERs), can be traded and are fully functional. The role of the EU ETS in the carbon market is very important, since a significant part of the demand for JI and CDM reduction units comes from the EU ETS sectors.[26]

According to Article 30 of the EATD, the Commission is required, on the basis of the experience of the implementation of the EU ETS and in relation to the developments in the international community on climate policy, to adopt a report on the application of the EATD. On 13 November 2006, the European Commission released this report.[27] The document sets the agenda for the revision of the EATD. The report focuses on four key aspects of the EATD, namely the scope, the further harmonization within the EU and the increased predictability of the market, a more robust compliance and enforcement system and, finally, the linking to ET schemes in third countries. Potential amendments to the EATD will eventually come into force as from 2013.

The cases before the ECJ and CFI in respect of the NAPs encompass the following issues:

- Case T-178/05, *United Kingdom* v. *Commission*[28]

In the first case on the EU ETS scheme, the Court of First Instance (CFI) concluded in favour of the United Kingdom that had filed a lawsuit against the European

---

[25] Directive 2004/101/EC, OJ 2004 L338, p. 18.

[26] Capoor, K. and Ambrosi, P. (2007). *State and Trends of the Carbon Market 2007*, Washington DC: Carbon Funds, World Bank.

[27] Communication COM(2006) 676 Building a global carbon market – Report pursuant to Article 30 of Directive 2003/87/EC (EATD).

[28] Judgment of the Court of First Instance of 23 November 2005 - United Kingdom v. Commission (Case T-178/05), OJ C 155, 25.6.2005.

Commission. The UK government requested the annulment of the Commission Decision of 12 April 2005 which rejected the adjustment to the UK NAP adding 20 million of EUAs to the British power sector. The CFI refused the position of the Commission that the environmental aims of the ETS would be undermined if countries were to be allowed to increase their allocations to domestic industry. The CFI ruled that amendments to the NAP submitted to the Commission are authorized until the adoption of its decision under Article 11(1) of the EATD. Furthermore, the CFI indicated the date of 30 September 2004 as a 'cut-off deadline' for amendments to the NAPs I. The Commission issued a new decision on 22 February 2006 rejecting once again the proposed amendment to the UK NAP, based on the fact after 30 September 2004 only amendments required by Commission decision are accepted. However, the UK government decided not to launch a court action against the Commission in order to preserve certainty over the level of EUAs for national installations falling under the scheme.

- Case T-374/04, *Germany* v. *Commission*[29]

The Commission decision of 7 July 2004 rejected the ex-post adjustments downwards to the NAP proposed by Germany on the basis of Criterion 5 of Annex III of the EATD, which states that 'the [national allocation] plan shall be consistent with other Community legislative and policy instruments. Account should be taken of unavoidable increases in emissions resulting from new legislative requirements'. The Commission claimed that the proposed ex-post adjustments would favor new entrants in place of existing operators. Germany requested the annulment of the Commission decision for three main reasons: 1) Annex III does not prohibit compensating over allocation with ex-post adjustments; 2) the Commission decision is in breach of Article 176 EC Treaty thus hindering more stringent measures for environmental protection; 3) new entrants would join the market with a disadvantaged position. No decision has yet been reached in this case by the time of completing this article.

- Case T-387/04, *EnBW Energy Baden Württemberg AG* v. *Commission*[30]

One of the main energy producers requested the annulment of the Commission decision of 7 July 2004 on the German NAP. EnBW Energy Baden Württemberg AG did not agree with the allocation methods for power stations decommissioning nuclear energy installations (such as one of its main competitors RWE) and considered this as state aid. EnBW claimed that the Commission failed to start the state aid procedure, breaching Article 88(2) of the EC Treaty. On 30 April 2007 the CFI decided that this request was inadmissible due to a lack of interest in bringing proceedings.[31]

---

[29] Action brought on 20 September 2004 by the Federal Republic of Germany against the Commission of the European Communities, (Case T-374/04), OJ C 284, 20.11.2004, p. 25.

[30] Action brought on 27 September 2004 by EnBW Energie Baden-Württemberg AG against the Commission of the European Communities, (Case T-387/04), OJ C 6, 08.01.2005, p. 38.

[31] N.y.r.

- Case T-32/07, *Slovakia* v. *Commission*

The Slovak government decided on 7 February 2007 to file a lawsuit against the European Commission for the annulment of the Commission decision of 29 November 2006 on the Slovak NAP for the phase 2008-2012. The Commission decision reduced the proposed total annual average quantity of allowances in Slovakia by 10.4 million EUAs on the ground that such a distribution is not consistent with the 'assessments made pursuant to Decision 280/2004/EC and not consistent with the potential, including the technological potential, of activities to reduce emissions [...] in addition, the part of the total quantity potentially amounting to 1.729389 million tonnes of allowances in respect of additional emissions of combustion installations annually to the extent that this is not justified in accordance with the general methodologies stated in the national allocation plan and on the basis of substantiated and verified emission figures'. It is argued that the Commission has no competence to dictate how the distribution of EUAs is to be calculated. The decision to challenge a NAP has no suspending effect on the Commission decision over the Slovak NAP and therefore this shall be implemented in the meanwhile.

- Case T-28/07, *Fels-Werke and others* v. *Commission*

Following the Commission decision of 29 November 2006 requiring a substantial cut in the proposed amount of EUAs by the German authorities and including the rejection of long-term allocation guarantees for new entrants, three German companies from the lime and glass industry have decided to start legal action before the CFI for the annulment of the mentioned Commission decision and to ask for the fast-track procedure. The German companies claim that 1) the decision was issued after the three-month deadline had expired; 2) the decision to cut the proposed quantity of EUAs was not justified with good reasons and that 3) the quantity of EUAs for new entrants does not constitute state aid under Article 87 of the EC Treaty. Furthermore, according to the three companies, the Commission decision 'is of direct and individual concern' to the plaintiff, thus providing *locus standi* under Article 230(4) of the EC Treaty. The Commission claims that the decision is addressed to Germany and has no direct effect for the plaintiff. The German government decided not to challenge the Commission decision.[32]

The difficulties encountered by the member states in the implementation of the EATD, in particular with reference to the definition of uniform and efficient national allocation plans, show that more stringent and precise guiding criteria for the preparation of these documents should be envisaged by the European Commission. The Commission should take the experience of the EU ETS into consideration when drafting the National Action Plans including sectoral targets and measures to meet the renewable energy targets as highlighted in the European Council Action Plan (2007-2009).[33]

---

[32] Information on this case was provided by the law firm Freshfield Bruckhaus Deringer which is currently dealing with the case.

[33] See Brussels European Council of 8/9 March 2007 Presidency Conclusions.

4. THE EU POSITION ON THE FUTURE CLIMATE POLICY

Article 3(9) of the Kyoto Protocol requires the Conference of the Parties acting as the Meeting of the Parties of the Kyoto Protocol (COP/MOP) to initiate the consideration of future commitments for Annex I Parties. The international talks over the post-2012 period officially started in 2006 in Bonn at the 24th session of the subsidiary bodies of the UNFCCC, where the first meeting of the Ad Hoc Working Group on Further Commitments for Annex I Parties under the Kyoto Protocol (AWG) – established at COP/MOP1 in Montreal from 28 November to 10 December 2005 – was held. The works of the AWG have to be reported to each COP/MOP and the decision should be adopted by the COP in time to ensure no break between the first and the second commitment period. At the first session of the AWG, the delegations agreed on the work plan and schedule for the establishment of new GHG emission reduction obligations for the post-2012 phase.[34] A decision on that matter is expected by COP/MOP3 in 2007.

European climate policy is increasingly focusing on future (post-2012) GHG emissions reduction strategy and targets. The Brussels European Council of 25-26 March 2004 officially opened the discussion on post-2012 actions towards climate change and the European heads of state and governments paved the way for the discussion over 'medium and longer term emission reduction strategies, including binding targets'. The EU policy towards post-2012 actions is based on the background paper Future Action on Climate Change: A Stakeholder Consultation on the EU's Contribution to Shaping the Future Global Climate Change Regime and on the Communication 'Winning the Battle Against Climate Change'.[35] The latter in particular focuses on the necessity to extend countries and sectors currently committed to the reduction of GHG emissions, to enhance the development of green technologies as well as to stick to flexible mechanisms and the promotion of adaptation measures. The initiative of the Commission was backed by the European Parliament's environment committee which on 27 April 2005 adopted the resolution 'Winning the Battle Against Global Climate Change',[36] highlighting the need for major efforts at the international negotiations on the post-2012 phase, indicating the possibility to establish trade sanctions for states not ratifying the Kyoto Protocol and to enhance green technologies on energy and transport.

Moreover, the Council Environment of 10 March 2005 started the discussion on the extent of new GHG reduction commitments after 2012 for the industrialized world and in particular for the EU member states. This was confirmed in the already mentioned energy package adopted by the European Commission in January 2007,[37] which included the reference to a commitment to reduce GHG emissions

---

[34] Ad Hoc Working Group on Further Commitments for Annex I Parties under the Kyoto Protocol, document FCCC/KP/AWG/2006/L.2/Rev.1.

[35] Communication of the European Commission COM(2005) 35 'Winning the Battle Against Climate Change', Brussels 9 February 2005.

[36] Document 2005/2049(INI).

[37] Communication from the Commission to the European Council and the European Parliament – An Energy Policy for Europe COM(2007) 1, Brussels, 10 January 2007.

by 20% by 2020 in respect of 1990 levels at the European level. The Energy Policy for Europe includes a list of proposed actions to be implemented by both developed and developing countries in order to limit global warming to no more than 2°C above pre-industrial temperatures.[38] The position of the Commission is supported by the Parliament[39] and by the EU member states as confirmed by the presidency conclusions of the European Council of March 2007.[40]

An interesting discussion which arises when looking at the new commitments for the reduction of GHG emissions in Europe in the period after 2012 concerns the distribution of the burden among member states. The current EU BSA considers only the old EU 15 and in accordance with Article 4(4) of the Kyoto Protocol no changes to this agreement can take effect before the end of the first commitment period. It is very likely that the EU composed of 27 member states will be aiming at a similar agreement for the post-2012 commitments at the international level, namely a joint reduction target to be distributed internally among the member states. This is confirmed by the Integrated Climate and Energy Policy (ICEP) included in the EU presidency conclusions of March 2007 which highlight the necessity for an agreed internal burden-sharing for the 27 member states based on national circumstances and on a differentiated approach. A few considerations need to be drawn in this respect. First, Malta and Cyprus, currently EU member states, are still not included in the list of Annex I parties to the UNFCCC and are therefore considered by the international climate regime as developing countries. These two states will need to have the same status as the rest of the EU countries once a joint commitment for the EU is negotiated. Moreover, a few new member states, namely Bulgaria, Hungary and Romania have already requested base years to be different from the common base year for the EU (1990) in the definition and negotiation of new binding targets for the reduction of GHG emissions. Finally, it is important to emphasize the double-track approach adopted by the EU in the definition of the targets identified under the mentioned EU presidency conclusions. Namely, the member states agreed to adopt a BSA based on differentiated commitments for the targets in the field of the promotion of renewable energy (20% by 2020) and the reduction of greenhouse gas emissions (-20% by 2020). No BSA approach has been agreed for targets in the promotion of energy efficiency (reduction of energy consumption through energy efficiency measures by 20% by 2020) and in the share of biofuels (10% in the EU transport fuel consumption by 2020). Here, all member states will have the same level of commitment.

---

[38] Communication Limiting Global Climate Change to 2 degrees Celsius – The way ahead for 2020 and beyond COM(2007) 2.

[39] In response to the Commission paper, on 14 February 2007 the Parliament adopted a resolution calling on the EU to reduce greenhouse gases by 20% by 2020 independently from the commitments of other industrialised countries.

[40] See n. 33 *supra*.

5. CONCLUSIONS

Although the latest information in the 2007 report on the level of GHG emissions in Europe by the European Environment Agency (EEA) showed that in 2005 the EU 15 GHG emissions dropped by 0.8% compared to 2004, the compliance of the old member states with the Kyoto Protocol target is still questionable.[41] EU 15 emissions of all GHG in 2007 are 1.5% lower than in 1990 (-8% is the target for the period 2008-2012).

The EEA confirmed that, according to the current trends in the GHG emission levels, the EC will be able to meet the 8% target agreed under the Kyoto Protocol only if all existing and planned domestic policy measures would be implemented, as well as the Kyoto mechanisms and the inclusion of carbon sinks.[42] The information on the GHG emission trends shows that while a few member states will probably miss their targets under the EU BSA, the compliance of the EC with the Kyoto Protocol may be ensured by the efforts of other member states and by the use of the flexible mechanisms. In this sense, still open is the question whether the non-compliance of a member state with the targets agreed under the EU BSA could lead to an infringement procedure under EC law for a failure to comply with European legislation by that member state. At the same time, EC competition law could be triggered in relation to the fact that some European companies may suffer from stricter decisions adopted by their national authorities in comparison with other member states who agreed on more lenient targets towards the Kyoto Protocol.

At the end of 2006, in accordance with Article 3 of Council Decision 2002/358/EC, the Commission submitted to the UNFCCC its GHG inventory for the period 1990-2004 including the final estimate of the EU 15 GHG emissions in 1990. On the basis of these data, the Commission established the 'emission levels allocated to the European Community and to each Member State in terms of tonnes of carbon dioxide equivalent' against which the EU 15 must reduce their emissions of 8% by 2008-2012.[43] According to the Commission estimations, the EU 15 will be allowed to emit an average of 4,278 $MtCO_2$ eq. per year in the Kyoto Protocol first commitment period. These data represent the EU final assigned amount units under the Kyoto Protocol.[44]

---

[41] EEA Technical report, Annual European Community greenhouse gas inventory 1990–2005 and inventory report 2007, July 2006.

[42] EEA Report No. 9/2006, Greenhouse gas emissions trends and projections in Europe 2006 and Report from the Commission on the Progress towards achieving the Kyoto objectives (required under Decision 280/2004/EC of the European Parliament and of the Council concerning a mechanism for monitoring Community greenhouse gas emissions and for implementing the Kyoto Protocol), COM(2006) 658, 27 October 2006.

[43] Commission Decision of 14 December 2006 determining the respective emission levels allocated to the Community and each of its member states under the Kyoto Protocol pursuant to Council Decision 2002/358/EC (2006/944/EC).

[44] Commission decision of 14 December 2006 determining the respective emission levels allocated to the Community and each of its member states under the Kyoto Protocol pursuant to Council Decision 2002/358/EC, OJ of the EU, L 358/87 of 16 December 2006.

The EU political commitment to maintain the lead in the international community in the fight against climate change, confirmed by the latest decision at the level of the European Council, has encountered a few delays in the transposition of policy actions and documents into legislation aimed at the reduction of GHG emissions. This is the case, for instance, concerning the adequacy and stringency of the NAPs developed under the EU ETS by the national authorities in the member states. In the first year, the externally-verified data on the CO_2 emissions registered in 2005 by the installations covered under the EU ETS, showed that the level of CO_2 emissions in the first year of emissions trading were 2.5% less than the EUAs received by the government through the NAPs. In the second phase, although the final version of the NAPs has been delayed in many cases by the member states, the Commission decided to reduce significantly the quantity of proposed EUAs allocated by the national authorities to their installations. This shows that there is still a certain leniency in the member states' decisions over the efforts required of the national industries in the adoption of mitigation measures for the reduction of GHG emissions.

Even more relevant is the case of CO_2 emissions from passenger cars. In this sector, it seems that the approach of the Commission based on the establishment of voluntary agreements with European and non-European car manufacturers' associations failed to meet the foreseen goals. Car manufacturers failed to meet their targets based on less than 140 CO_2 per g/km by 2008 for ACEA and by 2009 for JAMA and KAMA.[45] On the ground of this failure, in the 2007 strategy paper the Commission requested the reduction of CO_2 emissions from road transport and the introduction of binding targets for new cars and vans.

Another unresolved dispute within the EU concerns the introduction of strict and binding targets on the share of renewable energy for member states. This question goes back to 2004 when the European Commission tabled a proposal for future targets of renewable energy in the EU.[46] At that time, it was not possible to reach an agreement among the EU member states on this point and any decision on post-2010 targets for EU renewable electricity generation was postponed until 2007 by the Energy Council of 29 November 2004. The Commission green paper 'Energy Policy for Europe' includes Communication COM(2006) 848 which identifies a goal of a 20% share of renewable energy in the EU's energy mix by 2020.[47] Furthermore, on 14 February 2007 the European Parliament adopted a resolution on climate change including a proposition for a target of 25% of renewable energy in the share of EU energy use by 2020. Finally, the European Council of 8/9 March

---

[45] In 2006, CO_2 emissions from passenger cars amount to 163 g/km.

[46] Communication of the Commission to the Council and the European Parliament, The share of renewable energy in the EU Commission Report in accordance with Art. 3 of Directive 2001/77/EC, evaluation of the effect of legislative instruments and other Community policies on the development of the contribution of renewable energy sources in the EU and proposals for concrete actions COM (2004) 366 - Brussels, 26 May 2004.

[47] Communication Renewable Energy Road Map. Renewable energies in the 21st century: building a more sustainable future COM(2006) 848.

2007 agreed on binding targets for a 20% increase in the use of renewable energy by 2020 to be distributed in binding national targets for the 27 member states.

Meanwhile, a few member states have opted for proposals and actions aiming at the reduction of GHG emissions which even go beyond the measures tabled by the EU. In this respect, it can be pointed out that member states, in the field of the protection of the environment and in accordance with Article 176 EC Treaty, are not prevented from 'maintaining or introducing more stringent protective measures' than EC measures based on Article 175 EC Treaty. However, where EC measures are based on Article 95 EC Treaty – the internal market provision – strict conditions need to be met before more stringent national measures can be maintained or introduced. Also, the Commission needs to approve such stricter national measures. For instance, the European Commission decided in December 2006 to authorize Denmark to maintain limits on f-gases which are stricter than the obligations introduced by Regulation 842/2006/EC on certain industrial greenhouse gases. That regulation was based partly on Article 175, and partly on Article 95 EC Treaty. Due to the specific rules laid down in the Regulation, Denmark was allowed to maintain its stricter limits on f-gases until the year 2012.[48]

An example of an interesting proposal is the French idea of the introduction of an EU border-tax adjustment to protect the European market from any competitive advantage of imports from countries which do not have strict obligations in place on the reduction of GHG emissions. Trade Commissioner Peter Mandelson explained in December 2006 that he had serious doubts about such a specific 'climate' tariff on countries that have not ratified Kyoto. 'This would be highly problematic under current World Trade Organization (WTO) rules, and almost impossible to implement in practice.' The European Commission did not consider this idea in the Energy Policy for Europe package of early 2007.[49] Apparently, the legal challenges with regard to climate change inside the EU are considered to be already sufficient.

---

[48] Commission Decision of 8 December 2006 concerning national provisions notified by Denmark on certain industrial greenhouse gases, OJ of the EU of 6 February 2007, L 32/130.

[49] See IP/07/29 of 10 January 2007.

THE EU EMISSIONS TRADING DIRECTIVE: TIME FOR REVISION?

Nick Farnsworth*

1. THE FUTURE OF EMISSIONS TRADING: ADDITIONAL SECTORS AND GASES

The European Union Emissions Trading Scheme (EU ETS) is generally recognized for its pioneering steps to curb rising greenhouse gas emissions. However, the success of the scheme is still disputed. On the one hand, a new market in greenhouse gases has been created but so far it only covers carbon dioxide and only a handful of sectors, namely large emitters in the power and the heat generation industry and selected energy-intensive industry sectors, such as combustion plants, oil refineries, coke ovens, iron and steel plants and factories making cement, glass, lime, bricks, ceramics, pulp and paper. The environmental integrity of the scheme must be confirmed by expanding the scheme to cover more greenhouse gases, more sectors, and prove that trading is the most effective means of reducing emissions.

The first phase of the European Emissions trading scheme is often referred to as a 'learning-by-doing' stage. The European Environment Agency acknowledged that the scope of the EU ETS was intentionally limited during its initial phase while experience of emissions trading is built up.[1] Anticipating that changes would be needed to the EU ETS, the original EU ETS directive provided for a review by June 2006.[2] The review was only published in November 2006, in which the Commission confirms 'the first 18 months of the EU ETS has proven to be a very valuable learning period'. The Commission's review continues that 'for reasons of regulatory stability and predictability, any changes to the Directive emanating from this review should take effect at the start of the third trading period in 2013.' Regarding actual changes or amendments to the existing Directive, the Commission refrains from making any firm commitments for the third trading period and will continue to review the system in a Working Group of the European Climate Change Programme (ECCP).[3] On this basis, the actual review of the ETS is very cautious.

---

* ADS Insight, Bruxelles.
[1] EEA Technical report No. 2/2006 'Application of the emissions trading directive by EU Member States'.
[2] Communication from the Commission to the Council, the European Parliament, the European Economic and Social Committee and the Committee of the Regions, Building a global carbon market – report pursuant to Art. 30 of Directive 2003/87/EC COM(2006) 676.
[3] <http://ec.europa.eu/environment/climat/eccpii.htm>.

W.Th. Douma et al., eds., The Kyoto Protocol and Beyond
© 2007, T·M·C·ASSER PRESS, *The Hague, The Netherlands and the Authors*

Four issues dominate the review – 1) the scope of the Directive, 2) further harmonization and increased predictability, 3) robust compliance and enforcement and 4) links to third countries.

1.1 The scope of the directive

The EU ETS only covers about half of the industrial and energy sector. Greenhouse gases other than carbon dioxide, which are not covered by the scheme, still contribute up to 17% of overall emissions in the EU 25.[4] Nevertheless, there is little in Directive 2003/87/EC to prevent the EU ETS being extended to other greenhouse gases, after all the title of the Directive states that it is a scheme for 'greenhouse gas emissions' and not only 'carbon emissions'. However, as it stands, the scheme only allows for the trading of CO_2. This is because Annex I[5] limits trading in the European scheme to carbon dioxide. Other potential gases are listed in Annex II.[6] At least until 2013, methane (CH_4), nitrous oxide (N_2O) hydroflourocarbons (HFC_4), perflourocarbons (PFCs) and sulphur hexafluoride are excluded from trading activity within the European scheme. Nevertheless, the Linking Directive[7] allows member states to purchase 'all CERS and ERUs that are issued … in accordance with the UNFCCC and the Kyoto Protocol…'.

In the Commission's review, it states that the 'review will look at expanding the EU ETS to other sectors and gases including N_2O from the production of ammonia and CH_4 from coal mines.' These sectors emerged as the most suitable candidates to join following the LETS Update study[8] released in March 2005. To date, it is the most comprehensive analysis of the EU ETS by the Commission, but technically it is only a background paper and does not replace the review or the work of the European Climate Change Programme which will continue the formal review process.

---

[4] LETS Update, Decision Makers Summary, p. 11.

[5] Annex I – Categories of activities referred to in Arts. 2(1), 2, 4, 14(1), 28 and 30, Directive 2003/87/EC of the European Parliament and of the Council of 13 October 2003 establishing a scheme for Greenhouse Gas emission allowance trading within the Community and amending Council Directive 96/61/EC (the Emissions Trading Directive).

[6] Annex II – Greenhouse Gas referred to in Arts. 3 and 30, Directive 2003/87/EC of the European Parliament and of the Council of 13 October 2003 establishing a scheme for Greenhouse Gas emission allowance trading within the Community and amending Council Directive 96/61/EC (the Emissions Trading Directive).

[7] Directive 2004/101/EC of the European Parliament and of the Council amending Directive 2003/87/EC establishing a scheme for Greenhouse Gas emission allowance trading within the Community, in respect of the Kyoto Protocol's project mechanisms ('The Linking Directive').

[8] 'LIFE LETS Update: Decision Makers Summary', by LETS Update Advisory Group. The European Commission commissioned this comprehensive study to provide background as well as the necessary data for an update to the EU ETS. The study was prepared by the LETS Update Advisory Group which includes representatives of Climate Action Network Europe, the Centre for European Policy Studies (CEPSs), the Carbon Trust, the European Commission (DG Environment), the Ministry for Ecology (France), the Institute for Environmental Protection, Poland and the International Emissions Trading Association (IETA), and therefore represents a cross-section of business and private sector as well as leading NGOs, with an interest in climate change.

The study considered the criteria for extending the scheme, the volume of and trends in emissions, uncertainty levels in emissions, technical feasibility for reducing emissions, the number of and size of emitters and other relevant policies or regulations. Nitrous oxide (N_2O) in the chemical sector, perflourocarbons (PFCs) in aluminium production, methane (CH_4) from active coal mines and hydroflourocarbons (HFCs) in refrigeration and cooling were highlighted for more detailed study. The study also investigated expanding the list of activities to include more sectors emitting CO_2. After a screening process the authors identified the chemicals sector (fertilisers and ammonia, petrochemicals and other chemicals) and aluminium to be the most likely sectors for inclusion in future phases.

LETS Update was cautious about recommending the inclusion of the transport sector and, in particular, avoided discussion about the aviation sector. Nevertheless, in December 2006 the European Commission made a legislative proposal to include aviation.[9] The proposal has already been very controversial as it highlights several difficulties legislators face in broadening the scope of the EU ETS. The proposed Directive acknowledges that the greenhouse gas emissions from aviation have a wider impact than only CO_2. Airlines produce a cocktail of chemicals which together have a greater impact on the atmosphere, when released at altitude.

'Aviation has an impact on the global climate through release of CO_2, N_2O, water vapour and sulphate and soot particles. The International Panel on Climate Change (IPCC) has estimated that the total impact of aviation currently is two to four times higher than the effect of its CO_2 emissions alone. Recent Commission research indicates that the total impact of aviation could be around two times higher the impact of CO_2 alone. However, none of these estimates takes into account the highly uncertain cirrus cloud effects. Pending scientific progress to identify suitable metrics for comparing the different impacts, a pragmatic and precautionary approach is required. Emissions of nitrogen oxides will be addressed in other legislation to be presented by the Commission.'[10]

Whilst the Commission acknowledges the greater impact of aviation emissions, the proposed Directive to include aviation will only address CO_2 emissions. With so much scientific uncertainty regarding aviation emissions, it seems the Commission is clearly cautious not to push the limits of the precautionary principle although this could change if the forthcoming IPPC report, due in 2007, offers more detail about the effect of aviation on climate change.

It would be necessary to adapt the current EU ETS to integrate other gases but this should not be too difficult. If the EU ETS were broadened, it is likely that the EU ETS would apply similar rules to the flexible mechanism of the Kyoto Protocol. This would ensure consistency and make it easier to link the EU ETS with non-European schemes. One important scientific distinction made in the Kyoto Proto-

---

[9] Proposal for a Directive of the European Parliament and of the Council amending Directive 2003/87/EC so as to include aviation activities in the scheme for Greenhouse Gas emission allowance trading within the Community, 20/12/96 COM(2006) 818.

[10] Preamble 12 of the Proposed Directive.

col is that each greenhouse gas has very different properties, and consequently each gas has a different effect on the atmosphere. Under the flexible mechanism, greenhouse gases are scaled according to the global warming potential based on the Intergovernmental Panel on Climate Change (IPCC), which provides the generally accepted, though not undisputed values, for GWP. The values changed slightly between 1996 and 2001. Consequently, based on the current calculations, CO_2 has a GWP of one and all other gases which have a GWP have a carbon equivalent to CO_2, measured over a specific timescale (e.g., 100 years). For example, methane has a potential of 23 over 100 years.

Nevertheless, the EU ETS does not preclude member states from taking independent initiatives to integrate new sectors and gases; a principle recognized by the environmental provisions of the Treaty establishing the European Communities.[11]

'From 2008, Member States may apply emission allowance trading in accordance with [the] Directive to activities, installations and greenhouse gases which are not listed in Annex I, provided the inclusion of such activities, installations and Greenhouse Gases is approved by the Commission...'.[12]

Therefore even before 2012, member states may start to add gases, and sectors, on a national basis. Nevertheless such measures must be compatible with this Treaty and according to the directive, and measures shall be notified to the Commission. For example, in the Working Group of the European Climate Change Programme, France and the Netherlands have said they are considering options to include N_2O from adipic and nitric acid industries, whilst the UK, Lithuania and Finland are still considering their position.

Although there is great potential to expand the scope of the scheme to other sectors and gases, the review also highlights that the boundaries of the current scheme are not undisputed. One of the more controversial issues in the first reporting period regards combustion installations and the smallest installations. Member states have interpreted the guidelines widely leading to discrepancies whether the scheme is being applied consistently. In particular, it was uncertain whether the definition of combustion engines should include combustion installations for electricity, heat or steam production for energy production only, or generally. Furthermore, even if a Member state deems an installation to fall within the definition, there have been instances where installations can still opt out, for example in the Netherlands, or where an installation is subject to permitting regulations instead, such as in Germany. The review recognizes that clarity is needed. In an earlier review, the Centre for European Policy Studies (CEPS) suggested that Member states could agree together on a common definition; however, this would take time. Alternatively, the

---

[11] Art. 176 of the ECT 'The protective measures adopted pursuant to Article 175 shall not prevent any Member State from maintaining or introducing more stringent protective measures'.

[12] Art. 24(1) Directive 2003/87/EC of the European Parliament and of the Council of 13 October 2003 establishing a scheme for Greenhouse Gas emission allowance trading within the Community and amending Council Directive 96/61/EC (the Emissions Trading Directive).

Commission could amend Annex 1 of the directive and as a last resort, the Commission could initiate infringement proceedings against Member states.

Similarly, there has been concern that small installations are excessively penalized due to the high administration costs of accounting and monitoring. For small installations, the Commission will take into consideration the cost-effectiveness of including these installations.

1.2 Further harmonization and increased predictability

The review of the EU ETS acknowledges that initial allocation of allowances was the responsibility of each Member state. Whilst the Commission provided general guidance, the National Allocation Plans (NAPs) were the principle tool for each Member state to set out how many allowances to allocate for the specific trading period. Member states had the flexibility to set their cap, as well as how many allowances to allocate amongst the installations and potential new entrants. The Member state maintained control over their allocations so it was not surprising that National Allocation Plans varied dramatically, not only in structure and content but also the level of ambition. Already during the allocation for phase II, Member state allocations have been incredibly controversial as the Commission has forced several Member states to revise their overall cap downwards.

Arguably over-allocation has led to discrimination and an abuse of the internal market. It also abuses the fundamentals of the establishment of the market, namely a scarcity of emission allowances. An oversupply of emissions allowances would drive down the price of allowances. Although the first phase of NAPs went largely unchallenged by the Commission, to date no NAPs have formally been challenged on the basis of an abuse of State Aid or discrimination, perhaps because the baseline data was not adequate. Nevertheless in phase II, it is clear that the Commission is ready to take a tougher stance.

One obvious solution would be to centralize the allocation, thereby to avoid the possibility for Member states to discriminate in favor of their national industries. However, for a Member state to concede responsibility for allocation is extremely sensitive. Article 5 of the Treaty establishing the Community states that

'the Community shall take action, in accordance with the principle of subsidiarity, only if and in so far as the objectives of the proposed action cannot be sufficiently achieved by the Member States and can therefore, by reason of the scale or effects of the proposed action, be better achieved by the Community.'

The principles, known as the subsidiarity and proportionality principles, are elaborated in the 'Protocol on the application of the principles of subsidiarity and proportionality', Amsterdam Protocol 1997. Nevertheless, the Amsterdam Protocol acknowledge that non-discrimination is an area where Community action could be better than national action, in order to achieve the goals of the proposed action. There is therefore a strong argument to support a centralized allocation body to ensure that the market is not oversupplied with emissions allocations and to ensure that Member states do not favour certain industries in their territory.

Harmonization has been a recurring feature of ETS discussions, but in the review of the EU ETS, the Commission focuses on the need for harmonization of the rules for new entrants. The Commission states that a number of options exist to 'harmonize the treatment of new entrants and that having new entrants buy allowances in the market or in an auction is in accordance with the principle of equal treatment.' On the other hand, there are also strong arguments that only free allocations would be consistent with the rules applicable to other installations and free allocations would incentivize companies to invest in new technology installations, rather than continue to operate old installations which were allocated an excessive amount of emission allowances as a result of the grandfathering process. LETS Update suggests that developing benchmarking and investigating further growth rate scenarios for individual sectors would facilitate Community harmonization. In this respect the Commission's proposal to centralize allowance allocations for the aviation sector is consistent with LETS Update, even though it does discriminate the aviation sector from the existing installations.

On the face of it, the centralization of aviation emissions seems to discriminate against the aviation sector. Whilst the discriminatory principle, Article 13 of the Treaty establishing the European Communities, usually applies to sex, race, ethnic origin, religion or belief, disability, age or sexual orientation, arguably it could also apply to 'producers of competing goods'. There is no legal precedent, but it could be an interesting development of the 'non-discrimination principle' unless the Commission decides to harmonize the allocation procedure for all sectors in line with the proposal for aviation.

Whilst many groups seem to support the introduction of aviation into the EU ETS, the question how is more controversial. Not only would it be necessary to harmonize the inclusion of aviation into the EU ETS, but it must also be consistent with international rules for aviation, which, to date, has largely fallen under the remit of the International Civil Aviation Organisation or have been decided on a bilateral basis between individual countries. The Kyoto Protocol[13] obliges parties to pursue an international solution through the International Civil Aviation Organisation (ICAO). Most international airlines have adamantly defended that it is the right and the responsibility of ICAO to find a solution. After all, aviation is by nature international and therefore an international solution should be sought. The same principle applies to the maritime industry, but the complexity of integrating maritime issues into the EU ETS would be substantially different to aviation.

It is not surprising that the Commission defends the legal basis for the Directive in the preamble. The United States have indicated that they are not ruling out legal action, although the legal basis for doing so is unclear. The European Union maintains their right to impose protective measures, in much the same way as one country may impose road charges which drivers from other countries would be bound to

---

[13] Art. 6 Kyoto Protocol - 'pursue limitation or reduction of emissions of Greenhouse Gas from aviation bunker fuels, working through the ICAO'.

accept and pay. Nevertheless, the Commission quotes the sixth meeting of the ICAO Committee on Aviation Environmental Protection in 2004, where it was agreed that an aviation-specific emissions trading scheme based on a new legal instrument under ICAO auspices '... seemed sufficiently unattractive that it should not be pursued further...', and that 'Resolution 35-5 of the ICAO Assembly endorsed open emissions trading and requested the development of non-binding guidance for use by states, as appropriate, to incorporate emissions from international aviation into the ETS.' The development of the guidelines is ongoing and the European Commission and member states are involved in the process, which are scheduled for further discussion at the ICAO Assembly in September 2007. In the meantime, the European Commission has proposed that the new directive should include aviation and therefore it seems that the ICAO discussions will progress parallel to the co-decision procedure for the directive.

1.3   Robust compliance and enforcement

Compliance and enforcement procedures are a prerequisite for all legislation, not least for the EU ETS. Given the broad discretionary responsibility for member states to allocate emissions, it is important that legislation is also implemented consistently throughout all member states. An effective compliance and enforcement regime is also essential to facilitate the linking of the EU ETS with other trading schemes.

Currently the EU ETS is dependent on third party accreditation, although many stakeholders have stressed the need for more provisions at Community level. The Commission's review acknowledges that the Commission is considering the development of a Regulation to aid the harmonized application of the legislation. Furthermore, the DG Environment website[14] emphasizes that this is a UNFCCC and Kyoto Protocol requirement. The Commission acknowledges that

> 'it is appropriate for the European Commission to provide for effective cooperation and coordination in relation to the compilation of the Community greenhouse gas inventory, the evaluation of progress, the preparation of reports, as well as review and compliance procedures enabling the Community to comply with its reporting obligations under the Kyoto Protocol, as laid down in the political agreements and legal decisions taken at the seventh Conference of the Parties to the UNFCCC in Marrakech ("the Marrakech Accords").'

Furthermore,

> 'the objectives of complying with the Community's commitments under the Kyoto Protocol, in particular the monitoring and reporting requirements laid down therein, cannot, by their very nature, be sufficiently achieved by the Member States and can therefore be better achieved at Community level.'

---

[14] <http://ec.europa.eu/environment/climat/gge.htm>.

1.4 Links to third countries

The review acknowledged two ways that the EU ETS can be linked to third countries. First, the Commission will look at linking the scheme with other emissions trading schemes that are planned or are in operation. Such schemes could include the emissions trading schemes planned by the north-eastern States and California in the US or by Australian states.

In the Commission's proposal to include aviation, it is apparent that airlines from third countries will be obliged to participate. The proposed directive for aviation covers all flights arriving at or departing from an airport, i.e., European airlines and non-European airlines. The directive will treat all airlines equally, whether EU-based or foreign. From 2011 all domestic and international flights between EU airports will be covered, and from 2012 the scope will be extended to all international flights arriving at or departing from EU airports. The Preamble to the proposed directive states

> 'The community scheme can thereby serve as a model for the expansion of the scheme worldwide. If a third country adopts measures for reducing the climate impact of flights to a Community airport departing from that country which are at least equivalent to the requirements of this Directive, the scope of the Community scheme should be amended to exclude flights arriving in the Community from that country.'[15]

The Commission will also consider how to continue the development of the Kyoto Protocol's project-based mechanism beyond 2012. The Preamble to the Emissions Trading Directive provides that there are several benefits to opening up the EU ETS to JI and CDM. In Preamble 2, the Directive states that this 'will increase the cost-effectiveness of achieving reductions of global Greenhouse Gas'. Preamble 3 continues that project-based mechanisms to the Community scheme 'safeguard ... the environmental integrity ... increase the diversity of low-cost compliance options within the Community scheme ... and improve the liquidity of the Community market in Greenhouse Gas emissions allowances.' The directive also states that linking the EU ETS and Kyoto Protocol will 'stimulate' investment in the development and transfer of environmentally sound technologies and know-how.

In a similar vein, another key issue is how to enable developing countries to reduce their emissions. Implicitly, the Commission acknowledges that emissions trading and the Kyoto Protocol should not stand alone in terms of policy to help developing countries to reduce their emissions. DG Environment identified three ways in which developed countries, such as the European Union, can facilitate the transition for developing countries.[16] First, the Commission acknowledges that the Kyoto Protocol's Clean Development Mechanism (CDM) can be streamlined and expanded. The scope of the CDM could be expanded to cover entire national sec-

---

[15] Preamble 12 to the proposed Directive, see also proposed Art. 25a.
[16] Communication COM(2007) 2 'Limiting Global Climate Change to 2 degrees Celsius -The way ahead for 2020 and beyond'.

tors, generating emissions credits if the whole national sector exceeds a predefined emission standard. However, an expanded CDM can only function if there is increased demand for credits, and this will only happen if all developed countries take on substantial reduction obligations.

Second, the Commission acknowledged that developing countries need improved access to investment. Investment in new electricity generation in developing countries is projected to reach more than € 130 billion per year in order to support economic growth. The great majority of the resources will be generated by the major developing countries themselves. For example, reducing CO_2 emissions effectively in the power sector will require an additional investment of around € 25 billion per year. The Commission acknowledged that the gap cannot be filled through the CDM, even if expanded, or by development aid. Instead, it will require a combination of the CDM, development aid, innovative financing mechanisms (like the EU's Global Energy Efficiency and Renewable Energy Fund), targeted loans from international financial institutions and efforts by those developing countries that have the means. The earlier this gap can be filled, the less developing country emissions will grow.

Third, the Commission is considering how the introduction of sector-wide company-level emissions trading in sectors where the capacity exists to monitor emissions and ensure compliance, particularly for energy-intensive sectors such as power generation, aluminium, iron, steel, cement, refineries and pulp and paper, most of which are exposed to international competition. Through the World Business Council for Sustainable Development, The Cement Efficiency Workshop is considering how the cement industry can reduce emissions from cement manufacture at a global level. Such schemes would be either global or national; if national, schemes in developing countries should be linked with schemes in developed countries, with targets for each sector covered being gradually strengthened until they were similar to those set in developed countries. This would also limit the transfer of high-emission installations from countries where they are subject to reduction commitments to countries where they are not.

2. CONCLUSION

To conclude, the EU ETS remains the centerpiece of the European Climate Change Programme. The scheme seems certain to continue beyond 2012, but the Commission will refrain from further changes until the review is ready. In principle, there is opportunity to expand the scheme to include other sectors and gases without recourse to the cumbersome co-decision procedure that the aviation proposal will have to comply with. It would be possible to expand the scheme by making an amendment to Annex I through the comitology procedure, the committee system which oversees the acts implemented by the European Commission, as opposed to the co-decision procedure. In practical terms, the comitology procedure would mean that an amendment to extend the Directive to other gases in the EU ETS could be introduced in a couple of months as opposed to a couple of years. Obviously the comitology procedure is not a carte blanche for unlimited changes; any changes introduced would have to be approved by the committee, which includes representatives of all the member states.[17] Most significantly, it would not have to be approved by the Parliament, although members of the European Parliament would also be represented in the comitology process.

In the new energy policy for Europe[18] released on 10 January 2007 the Commission emphasizes the need for the review to consider how to strengthen the EU ETS. Specifically, the Commission says that it would like to extend the scheme to other gases and sectors. Other key areas the Commission will look at include extending the allocation periods, recognizing carbon capture and storage, harmonizing the allocation process and linking the scheme with other mandatory schemes. As far as the proposal for aviation is concerned, centralizing the allocation process and obliging non-member States to join the European scheme seem to be more consistent with the new energy policy than the EU ETS as it currently stands and it seems this gives a clear indication in which direction the Commission would like to pursue the future of the EU ETS.

---

[17] Art. 23, Directive 2003/87/EC of the European Parliament and of the Council of 13 October 2003 establishing a scheme for Greenhouse Gas emission allowance trading within the Community and amending Council Directive 96/61/EC (the Emissions Trading Directive).

[18] Communication from the Commission to the Council, the European Parliament, the European Economic and Social Committee and the Committee of the Regions 'Limiting Global Climate Change to 2 degrees Celsius - The way ahead for 2020 and beyond' COM(2007) 2.

THE ROLE OF JOINT IMPLEMENTATION WITHIN THE CONTEXT OF EU POLICIES

Wytze van der Gaast*

1. INTRODUCTION

The concept of Joint Implementation (JI) has been part of global climate policy making since its inclusion in the United Nations Framework Convention on Climate Change (UNFCCC) of May 1992. Article 4.2(a) of the Convention states that:

> ' ...developed country Parties and other Parties included in Annex I may *implement* ... policies and measures *jointly* with other Parties and may assist other Parties in contributing to the achievement of the objective of the Convention.'

This Article provides the basis for JI. It allows an Annex I Party (i.e., an industrialized country) to fulfill its GHG emission reduction or limitation targets via abatement actions in cooperation with and on the territory of another Party. Such cooperation would involve a bilateral or multilateral agreement in which countries with high (marginal) costs of greenhouse gas (GHG) emission reduction invest in abatement projects in countries with relatively low (marginal) costs. The first country, the investor, would receive a credit for the resulting emission reduction.

Soon after the adoption of the UNFCCC, a controversy arose among countries on whether it was fair to allow industrialized countries to implement relatively cheap GHG abatement measures on the territory of, e.g., developing countries. Sceptics argued that JI could divert attention away from domestic investments in sustainable energy systems in industrialized countries. JI proponents, on the other hand, argued that the mechanism would increase the cost-effectiveness of global climate change action. In addition, proponents argued that well-designed JI projects would stimulate a transfer of (private) capital and technologies from industrialized countries to developing countries or to countries with economies in transition, which would otherwise not have taken place.

Since December 1997, JI has been part of the Kyoto Protocol as a project-based flexibility mechanism with which industrialized countries can fulfill their quantified emission reduction and limitation commitments partly on the territory of other industrialized countries. For similar project cooperation with developing countries, the Clean Development Mechanism (CDM) has been established.

---

* Joint Implementation Network (JIN) Foundation, Paterswolde, the Netherlands.

W.Th. Douma et al., eds., The Kyoto Protocol and Beyond
© 2007, T·M·C·ASSER PRESS, *The Hague, The Netherlands and the Authors*

This paper describes the present state of play concerning JI and discusses its compatibility with the EU accession of a number of Central and Eastern European countries (including compliance with the *Acquis Communautaire* and the EU emissions trading scheme).

2. STATE OF PLAY CONCERNING JI

When JI appeared in the UNFCCC, it was not a new concept. Earlier, some forms of cooperation among countries similar to JI had been introduced in other international agreements, treaties, and conventions, such as in the *Convention Concerning the Protection of the Rhine Against Pollution by Chlorides* (1976), the *UN Economic Commission for Europe Convention on Long-Range Transboundary Air Pollution* (1979), the *Vienna Convention on the Protection of the Ozone Layer* (1985), and the *Montreal Protocol to the Vienna Convention* (1987).[1] Each of these conventions enabled Parties to cooperate with other Parties in order to achieve targets at lower overall costs.

However, the cost-effectiveness potential in the context of climate change policies is potentially (much) larger than in the more region-based conventions mentioned above. For example, JI in the *ECE Convention on Long-Range Transboundary Air Pollution* has been complicated by the fact that most of the gases covered by the ECE Convention (e.g., sulphur dioxide) do not mix evenly in the atmosphere, but deposit, after their emission, in a region which is relatively near to the source of pollution. This situation has often been referred to as the SO_2 dilemma: the extent to which JI can be applied is limited due to the regional disposition of the pollutant.

Despite the inclusion of JI in the UNFCCC, Parties could not agree on how to apply the mechanism in industrialized countries' efforts to stabilize their GHG emissions at 1990 levels. Consequently, when the first Conference of the Parties to the UNFCCC (COP-1) was held in 1995, it was decided to establish a pilot phase for JI. During this phase, Parties could experiment with the JI concept, including accounting methodologies for calculating the GHG emission reductions.

The inclusion of JI in the Kyoto Protocol as a mechanism for project-based cooperation among industrialized countries (defined in Article 6 of the Protocol) was followed by a set of modalities and procedures for JI adopted at COP-7 in 2001 (held in Marrakech, Morocco). The main accounting principle of JI is that the emission reductions achieved by a project are transferred as emission reduction units (ERUs) from the country where the project takes place (the host country) to the country which buys these ERUs. The ERUs are deducted from the host country's emissions budget under the Kyoto Protocol (assigned amount) and added to the budget of the investor country. As the emission budgets have only been assigned to Annex I Parties during 2008-2012, JI projects can only result in ERU transfers during 2008-2012.

---

[1] Schrijver, 1995, in: Jepma, 1995, pp. 133-134.

However, this does not imply that countries could not start JI projects before 2008. In fact, several projects have already been implemented since the year 2000 (see below) and the emission reductions are traded among the participating countries either as ERUs (for emission reductions between 2008 and 2012), or assigned amount units (for emission reductions before 2005). The latter transfer is possible in case the host country knows that its GHG emissions budget is higher than its actual and/or expected GHG emissions. Pre-2008 JI projects could then be used to sell these emission surpluses to other industrialized countries.

At the time of writing (March 2007), the pipeline of JI projects consists of 155 activities, which are in the process of validation of the project plan or already in the process of determination. The latter refers to the decision of the JI Supervisory Board (JISC, the UN body established to supervise JI) on whether the project is in accordance with the modalities and procedures for JI as determined by the first Meeting of the Parties to the Kyoto Protocol (COP-MOP1) in December 2005 (Montreal).[2] Figure 2.1 shows the division of JI projects when expressed in terms of activities per project category. It becomes clear from the Figure that most of the present JI projects are in the category of renewable energy production, followed by energy efficiency projects and methane emission reduction projects at landfills, coalbeds and coalmines.

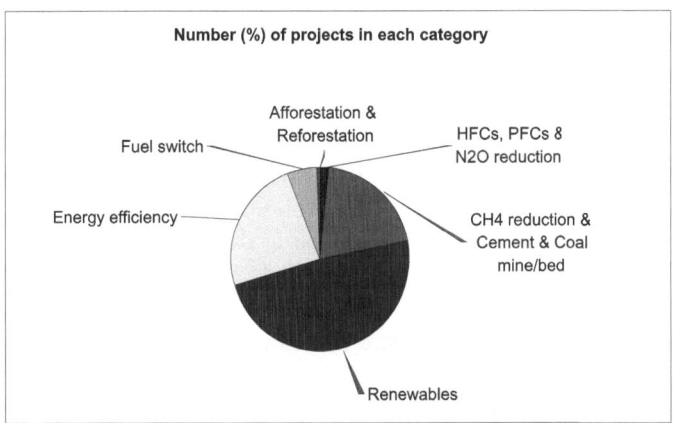

Figure 2.1. Number of JI projects in terms of project categories. *Source*: Fenhann (2007)

When looking at the project pipeline in terms of how many ERU projects are estimated to be delivered up to 2012, the picture changes with respect to the share of energy efficiency improvement and renewable energy production. Generally, renewable energy projects are relatively small in terms of scale, whereas energy efficiency projects, in particular in Central and Eastern Europe, could be carried out in large-scale industrial processes and district heating systems. In comparison to the

<hr>

[2] COP-MOP-1 took place after the entry into force of the Kyoto Protocol in February 2005 and was held in conjunction with the eleventh meeting of the Conference of the Parties to the UNFCCC.

CDM project pipeline, it can be noted that both in terms of projects per category and ERUs per project type, energy efficiency projects have a larger share in the JI pipeline. A possible explanation for this is that in most developing countries district heating systems are absent, so that energy efficiency under the CDM is mainly applicable in industrial processes.

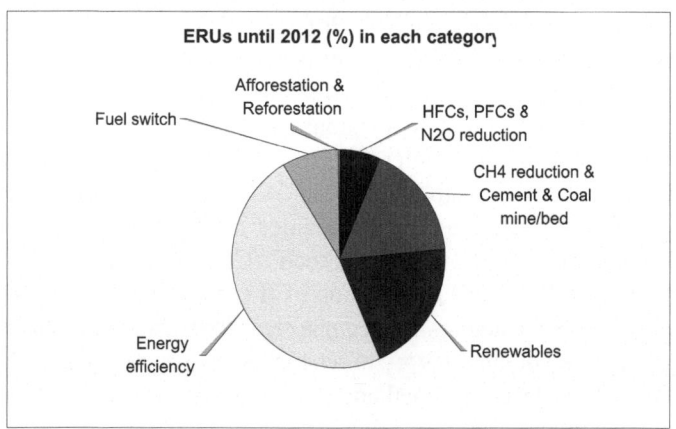

Figure 2.2. Number of JI projects in terms of ERUs. *Source*: Fenhann (2007)

Since its inclusion in the Kyoto Protocol, JI has been subject to a changing political environment. Firstly, ten countries from Central and Eastern Europe, which were initially expected to become important host countries for JI projects, have become EU member states in May 2004 and January 2007 (together with Malta and Cyprus). During the process of EU accession, these countries were required to comply with the EU environmental standards as included in the accession documents (*Acquis Communautaire*). By improving their standards to EU levels, energy efficiency improvement and GHG emission reduction efforts, which could otherwise have been undertaken as JI projects, became compulsory. Moreover, due to the EU accession the large emitters of GHGs in these countries have become subject to the EU Emissions Trading Scheme (EU ETS).[3] Emission reductions by these emitters are presently traded within the EU ETS, instead of through JI projects.

Within the pipeline, the Netherlands Government is involved as an investor in most of the projects: 37. Denmark, the World Bank (through its carbon procurement facility) and Austria follow with 28, 18 and 14 project activities, respectively. The Netherlands' portfolio of JI projects has been put together through a number of acquisition paths of which the ERU Procurement Tender (ERUPT) is the most important one. It was launched in May 2000 with the aim to invite international project developers to submit project ideas. The tenders were held annually until 2005. After that it was decided that ERUPT would continue in a new, more flexible style in which project developers no longer have to wait for tender announcements, but can

---

[3] JIN, 2003.

Table 2.1. ERU acquisition by the Netherlands Government

Acquisition track	Number of projects
ERUPT 0	1
ERUPT 1	3
ERUPT 2	0
ERUPT 3	3
ERUPT 4	10
ERUPT 5	10
ERUPT new style	2
Netherlands-EBRD carbon fund	4
Netherlands-EBRD + NEFCO	1
Consultants	1
Netherlands-IBRD	3
Total	**38**

Source: Fenhann, 2007.

immediately contact the Netherlands Government with project proposals. Table 2.1 shows how the contracted projects are spread across the ERU acquisition tracks.

Table 2.2 provides an overview of JI projects categorized per host country. It shows that the Russian Federation has become the largest supplier of JI projects. Eight of the current Russian JI activities are in the field of fugitive methane capture from gas distribution networks, six projects are in the category of producing bio-mass energy from waste paper and wood, and five projects envisage a switch from coal to gas in energy production. The remaining one-third of the Russian JI pipeline consists of small numbers of biogas, coalbed methane, energy efficiency in indus-try, the supply side and distribution, and landfill gas management. The Czech Re-public has a large number of projects (17) in the field of hydropower, which have been set up in cooperation with the World Bank Carbon Facility and are largely

Table 2.2. Division of projects across host countries

Russian Federation	31
Czech Republic	21
Bulgaria	20
Ukraine	17
Romania	15
Poland	13
Hungary	11
Estonia	11
Lithuania	5
New Zealand	5
Slovakia	3
Germany	3
Total	155

Source: Fenhann, 2007.

extensions of existing dams.[4] In terms of other project categories, the Czech Republic has undertaken little JI activity. Other important suppliers of JI projects are Romania and Bulgaria, which offer several energy efficiency, landfill gas capture and renewable energy projects. The latter two countries have received considerable technical support from the Netherlands Government (which started in 2001) to establish the capacity for JI project identification, approval and governance.[5] Both countries were among the first to sign Memoranda of Understanding with the Netherlands Government (Romania, November 1999 with an extension in November 2000; Bulgaria, April 2000)[6] and have been active in the Dutch ERUPT rounds.

On 28 March 2007, JISC for the first time determined a JI project. Determination in this context means that the project design document has been validated and that the subsequent round of public comments has not resulted in objections to the project. It implies that the project has become an official JI activity and can start generating ERUs (which, of course, will be subject to monitoring and verification). The determined project aims at a switch from wet to dry cement production in the Podilsky cement factory in Ukraine. By doing so, the project is expected to improve energy efficiency by 53%, which will deliver an annual CO_2 emission reduction of 750 ktonnes. Given that the project will start delivering ERUs as of 1 January 2009, it is expected to generate approximately 3 million ERUs (1 ERU = 1 tonne CO_2 emission reduction) up until the end of the Kyoto Protocol commitment period in 2012.

3.   COMPATIBILITY OF JI WITH EU POLICIES

As was explained above, the potential supply of JI projects within Europe has been reduced due to the accession of Central and Eastern European countries to the EU in May 2004 and January 2007. This has had two main implications. First, countries had to comply with the standards of the *Acquis Communautaire* and, second, large-scale GHG emitters from these countries have been assigned with emission caps under the EU ETS. These two compatibility aspects are discussed below.

3.1   JI and the *Acquis Communautaire*

In 1998, shortly after the adoption of the Kyoto Protocol, Central and Eastern European countries started a negotiation process with the European Commission, which would eventually lead to their accession to the EU.[7] In order to become eligible for

---

[4] Fenhann, 2007.

[5] JIN, 2001.

[6] JIN, 2000; JIN, 2001.

[7] The countries in Central and Eastern Europe which acceded to the EU in May 2004 and January 2007 are: Bulgaria, the Czech Republic, Estonia, Hungary, Latvia, Lithuania, Poland, Romania, Slovakia and Slovenia. Other new member states in May 2004 were Cyprus and Malta, but since these two countries are not included in Annex I of the UNFCCC, they are not eligible for JI project cooperation. Therefore, these countries have been left out of the scope of this paper.

EU Membership, the countries had to harmonize their national laws with EU legislation. The *Acquis Communautaire* represents a common level of requirements in all fields of EC law, which apply to all EU members as a safeguard to protect the European common market. The pre-accession process therefore involves harmonizing the national law in candidate countries with the common legislation in the EU reflected in the *Acquis*.

Part of the *Acquis Communautaire* is the set of common environmental EU standards, mostly in the form of directives which aim to protect environmental quality, to prevent pollution and to adjust production processes in order to make them more energy efficient.[8] Priority areas addressed by the EU environmental *Acquis* are: nature and biodiversity, the environment and health, natural resources, waste, water, air pollution and climate change.

Generally, it can be argued that the scope for JI projects has been reduced by the environmental standards of the *Acquis*. The incorporation of EU rules into the national legal systems leads to an improvement of the environmental standards in the new member states which contributes to the reduction of GHG emissions. Due to the *Acquis* environmental standards, a country's business-as-usual emissions level from which the JI project baselines are to be derived becomes (much) lower, thereby lowering the JI emission reduction potential. In other words, part of the potential JI abatement is already obligatory under the *Acquis Communautaire*. The implications of this effect may differ from country to country, as it depends, for instance, on the transition periods that states have agreed with the European Commission (i.e., in some cases states may incorporate some of the *Acquis* measures at a later date in order to smooth out the transition towards EU membership) and on the definition of *best available techniques* in the new member states (see below).

The example given in Nondek, et al. (2001) for the Czech Republic is illustrative in this respect. They calculated the technical GHG emission reduction potential in the year 2010, i.e., emission reduction measures beyond the requirements adopted in the Czech Republic between 1995 and 2000. This potential amounts to a reduction of 62 Mt CO_2-eq. for 2010. Next, adjusting this technical potential with economic effectiveness indicators (e.g., different levels of economic viability) results, according to Nondek, et al. (2001), in a 'real potential' of 24.5 Mt CO_2-eq. emission reduction in the Czech Republic in the same year. Assuming that part of this potential will by 2010 have been carried out anyway because of the economic attractiveness of some of the options and/or because of the requirements of the *Acquis Communautaire*, there is a 'real JI potential' result of 13 Mt CO_2-eq.

Although it is required that the new member states' environmental legislation is fully harmonized with EU legislation, there are still reasons why this does not auto-

---

[8] See, for instance, Council Directive 93/76/EEC of 13 September 1993 to limit CO_2 emissions by improving energy efficiency (SAVE) (European Council, 1993), Council Directive 1999/31/EC of 26 April 1999 on the landfill of waste (European Council, 1999), the Large Combustion Plant Directive 2001/80/EC (European Commission, 2001), and Directive 2006/32/EC of the European Parliament and of the Council of 5 April 2006 on energy end-use efficiency and energy services and repealing Council Directive 93/76/EEC.

matically imply that by the time a candidate country becomes an EU member, its environmental protection standards are exactly the same as the common standards of the EU.

Firstly, the country could agree with the EU on a timeframe for implementing the *Acquis* standards, which allows it to complete the full implementation of the *Acquis* measures later.[9] Only where measures require a substantial adaptation of the infrastructure in the candidate country, which needs to be spread over time (e.g., upgrading existing power plants to the IPPC Directive standards[10]) can a transition period be considered. The European Commission applies in this context the ground rule that the transition periods allow candidates to deal with the legacy of the past, but not to attract new investments with lower environmental standards.

The timing of implementing *Acquis* measures could have a significant impact on the potential for JI in the region. Suppose a country is expected to become a member of the EU in 2008 and negotiates transition periods of 5 years for some fields of the environmental *Acquis*, so that these have to be implemented domestically in 2013. For the entire first commitment period of the Kyoto Protocol 2008-2012, the baselines for the JI projects in this country would be derived from the 'Pre-accession Business-as-usual situation' instead of from the *Acquis* emission levels.[11] Countries that became EU member state in 2004 or shortly thereafter are, however, expected (if not required) to have most of the *Acquis* standards implemented in their domestic legislation by 2008. This implies that these countries' JI project baselines must be derived from the emissions scenario including the *Acquis*, which will lower the emission reduction potential of the projects.

Secondly, EU environmental legislation is often defined in terms of so-called 'Best Available Techniques' standards, i.e., each EU member state should introduce domestically the best available technology (BAT) for environmental protection. Conceptually, the EU defines BAT as:

– The most effective for the environment as a whole ('Best'),
– Already developed and possible to implement under economically and technically viable conditions ('Available'), and
– Including equipment and operational practice ('Techniques').[12]

---

[9] Such transition periods are not unique for the environmental chapter of the *Acquis Communautaire*. For instance, Poland has requested from the European Commission a transition period of 18 years with respect to fully opening up its real estate market to potential buyers from the EU. By the end of 2001, the European Commission had offered Poland a transition period of 11 years.

[10] IPPC stands for Integrated Pollution Prevention and Control and it contains environmental standards laid down in EU Council Directives.

[11] It should be noted, however, that the coming into force of *Acquis* standards is mostly a gradual process rather than a sudden change in legislation. Therefore, if for a country the '*Acquis* deadline' is 2013, part of the legislation must already have been incorporated before that date.

[12] Techniques in this context refer to, e.g., maintenance, choice of raw materials, management practices, monitoring procedures, etc. So, the BAT concept does not only refer to technologies.

It may be clear from the above that it is crucial how 'Available' is defined. According to the IPPC Directive, it means techniques

' ...*[d]eveloped on a scale which allows implementation in the relevant industrial sector, under economically and technically viable conditions, taking into consideration the costs and advantages ... as long as they are reasonably accessible to the operator* (emphasis added).' [13]

From this definition it can be concluded that 'available' is a relative term, which implies that 'Best Available Techniques' can differ between countries, but also within the countries. An important implication of the above definition is that interpreting and defining 'Best Available Techniques' for environmental standards, to a certain extent, has been a topic of negotiation between the EU and the candidate countries.[14] A result of this negotiation could be that for one new member state the BAT will become less strict than for another country. Therefore, a strict BAT indirectly reduces the potential for acquiring JI credits in the country since the improvement of the environmental standards are likely to reduce the emissions of GHGs. The opposite holds true for a less strict BAT.

3.2 JI and the EU Emissions Trading Scheme

Since January 2005, within the EU a Community-wide CO_2 emissions trading scheme has been operational. The scheme places caps on the emissions that installations in Europe may have each year. In order to remain below their caps, installations can either adjust their production and/or energy consumption patterns, or, in case their emissions surpass the caps, purchase emission allowances from other installations that have a surplus below their annual caps. The installations covered by the scheme are: combustion installations with a rated thermal input exceeding 20 MW, mineral oil refineries, coke ovens, installations producing and processing ferrous metals, mineral industry installations (cement clinker, glass and ceramic bricks), and pulp, paper and board producing and processing installations.

The EU Emissions Trading Scheme (EU ETS) is based on Directive 2003/87/EC.[15] The EU ETS runs for two phases: the first one is a learning, though legally-binding, phase which takes place during 2005-2007, and the second phase coincides with the 2008-2012 commitment period of the Kyoto Protocol. In the first phase, approximately 12,000 installations from 25 EU member states (including the 10 new member states of May 2004) have been allocated an annual CO_2 emissions budget by the governments of the member states. This allocation takes place

---

[13] EU Council Directive 96/61/EC of 24 September 1996 concerning the Integrated Pollution Prevention and Control (IPPC).

[14] The fact that the emission standards are negotiated with individual plant operators could imply that the plants' maximum emissions vary from plant to plant.

[15] Directive 2003/87/EC of the European Parliament and of the Council of 13 October 2003 establishing a scheme for GHG emission allowance trading within the Community and amending Council Directive 96/61/EC.

through so-called National Allocation Plans (NAP) which member states must prepare before an EU ETS trading phase and which is subject to the judgment of the European Commission.

Although the EU ETS has been designed as a stand-alone policy instrument – i.e., independently, before the entry into force of the Kyoto Protocol – it can be considered an important instrument for the EU and its member states to comply with the Kyoto Protocol commitments. The EU ETS is further linked to the Kyoto Protocol by the so-called Linking Directive 2004/101/EC, which allows EU installations to purchase emission reduction credits from abroad through JI and the CDM.[16] Under the Linking Directive, emission reductions achieved through JI projects can be traded on the EU market as of 2008 (during the second phase), whereas CDM credits have already been eligible for trade within the EU ETS as of 2005. The main reason for this difference is that JI projects, as explained above, can only generate ERUs during the commitment period of the Kyoto Protocol (2008-2012), whereas CDM credits could already be certified and banked for later compliance since 2000.

The linking of JI/CDM to the EU ETS takes place based on a one-to-one credit conversion rate, which implies that one tonne of CO_2-equivalent emission reduction achieved through a JI or CDM project and certified by the UN bodies created under the Kyoto Protocol for the governance of the CDM and JI (i.e., the CDM Executive Board and the JI Supervisory Committee) can be transferred to the EU ETS for the value of one European emission allowance – EUA (equivalent to one tonne of CO_2). Despite the choice of this equal conversion rate, the EU ETS and JI/CDM project mechanisms are fundamentally different systems. EUAs are allocated to EU installations and are thus free of uncertainties, whereas the GHG emission reductions generated by JI/CDM projects are calculated as reductions below a hypothetical baseline emissions scenario. By deciding on a one-to-one conversion rate of JI/CDM credits into EU allowances, the EU Council expressed its confidence in the credibility of the CDM accounting procedures, as supervised by the EB.

Originally, the European Commission proposed that the extent to which JI/CDM credits could be used for compliance with EU ETS should be limited to 6% of the total amount of allowances. In other words, an installation with 100 tonnes of allowances but with 110 tonnes of actual emissions could purchase 6 tonnes of JI/CDM credits and would have to purchase 4 tonnes on the EU market. However, in the Linking Directive no limitation to the transfer of JI/CDM credits to the EU ETS was included. Instead, the European Parliament required the inclusion in the final text of the directive of a reference to the fact that each EU member state shall decide on limitations to JI/CDM credit purchase for ETS compliance at the installation level.[17]

---

[16] Directive 2004/101/EC of the European Parliament and of the Council of 27 October 2004 amending Directive 2003/87/EC establishing a scheme for greenhouse gas emission allowance trading within the Community, in respect of the Kyoto Protocol's project mechanisms (Linking Directive).

[17] Linking Directive, para. 5, p. 1

In the course of 2006, EU member states submitted their NAPs for the trading period of 2008-2012, in which they also proposed caps on the use of JI/CDM. The final decisions adopted by the European Commission on the JI/CDM cap proposals are shown in Table 3.1.[18]

Tabel 3.1. CO_2 emission allowances in second NAPs and possible use of JI/CDM credits

	Average annual allowances 2008-2012 in NAP-II ($MtCO_2e$)	Limit to use of JI/CDM credits as per decision European Commission (% of total allowances)	Proposed JI/CDM credit limits subject to European Commission judgement (April 2007
Austria	30,7	10%	
Belgium	58,8	4-24%	
Bulgaria	56,2		20%
Cyprus	7,12		-
Czech Republic	86,8	10%	
Denmark	24,5		19%
Estonia	24,5		0%
Finland	39,6		12%-35,4%
France	132,8	13,5	
Germany	453,1	12%	
Greece	69,1	9%	
Hungary	30,8		10%
Ireland	21,2	34%	
Italy	209	9%	
Latvia	3,3	10%	
Lithuania	8,8	13%	
Luxembourg	2,7	10%	
Malta	2,1	-	
Netherlands	85,5	10%	
Poland	208,5	10%	
Portugal	37,9		10%
Romania	97,6		10%
Slovakia	30,9	7%	
Slovenia	8,3	17,5%	
Spain	152,7		20%
Sweden	22,8	10%	
UK	246,2	8%	
Total	**2127,0**		

Source: Second National Allocation Plans and European Commission decisions on allocation plans during December 2006 – March 2007 (a number of allocation plans are still the subject of European Commission considerations).

From the above it can be concluded that the effect of the EU ETS on JI could be twofold. On the one hand, the Linking Directive states that

'in order to avoid double counting, CERs and ERUs should not be issued as a result of project activities undertaken within the Community that also lead to a reduction in, or limitation of, emissions from installations covered by Directive 2003/87/EC, unless an

---

[18] At the time of writing for some member states no decisions have yet been taken by the European Commission.

equal number of allowances is cancelled from the registry of the Member State of the CERs' or ERUs' origin.'[19]

The reason for this decision by the European Parliament and the European Council is straightforward. Should an EU ETS installation become engaged in a JI project and sell ERUs to a foreign investor who uses these for Kyoto Protocol compliance, the emission reductions would also appear in the installation's EU ETS emissions bookkeeping. This would imply a double counting of the same emission reduction. Consequently, as per the Linking Directive, a project leading to emission reductions within the EU ETS cannot, in principle, generate ERUs or CERs for compliance with Kyoto Protocol commitments.[20] In case double counting is unavoidable, e.g., because an installation has already become engaged in a JI project before the country's EU accession, the member state could reserve in its NAP a so-called 'JI or CDM set-aside'. The double-counted emission reductions are then subtracted from this national NAP reserve.

On the other hand, since the Linking Directive allows EU installations to purchase JI and CDM credits for compliance with the EU ETS, demand for JI and CDM projects could increase. Before the EU ETS, JI and CDM projects were only implemented with a view to industrialized countries' compliance with Kyoto Protocol commitments. Now, with the EU ETS also extra demand for JI and CDM projects could come from the EU private sector. The main reason for adopting the Linking Directive was to

'increase the diversity of low-cost compliance options within the Community scheme leading to a reduction of the overall costs of compliance with the Kyoto Protocol while improving the liquidity of the Community market in greenhouse gas emission allowances.'[21]

In other words, the European Parliament and European Council were hoping that linking the EU ETS with JI and the CDM would create arbitrage opportunities, i.e., that ERUs and CERs would become cheaper than EUAs and that it would be more cost-effective to invest in JI/CDM projects than to purchase EUAs within the EU ETS market.

At the time of writing, however, the EU ETS prices are lower than the average price of CERs and ERUs: around € 0.50 for an EUA due for the first period € 5 and € 10 per tonne CO_2-eq. for JI and CDM credits. Initially, in the course of 2005, EU ETS prices strongly increased as several European installations immediately responded to GHG emission growths by purchasing extra allowances. Consequently, sharp oil price increases and dry summer conditions in Southern Europe immedi-

---

[19] Linking Directive, 2004, para. 10, p. 2.

[20] Note that the Linking Directive, in its para. 10, refers to ERUs and CERs. The latter credit is referred to because of the possibility that projects in Malta and Cyprus could be subject to double counting. Since Malta and Cyprus are not included in Annex I to the UNFCCC, projects in these countries fall under the Clean Development Mechanism of the Kyoto Protocol.

[21] Linking Directive, 2004, para. 3, p. 1.

ately led to EU ETS price increases. Eventually, prices increased to a level of over € 30 per tonne CO_2.

In May 2006, however, it turned out that overall CO_2 emissions in the EU ETS during 2005 had remained below the allocated allowances. After that, prices strongly dropped. With present prices, there are no real arbitrage opportunities between JI and CDM on the one hand, and the EU ETS on the other, because ERUs and CERs are presently more expensive than EUAs. This could change, though, during the second ETS period given that the present forward market price of EU allowances, tradable in December 2008, is close to € 20 per tonne (as per April 2007).

4. ENHANCING THE SCOPE FOR JI THROUGH STANDARDIZATION OF ACCOUNTING PROCEDURES

4.1 Introduction

Given the reduced scope for JI in the new member states, due to the EU ETS and the *Acquis Communautaire*, the main potential for JI in these countries seems to exist in those activities that are not covered by the EU ETS (including activities reducing greenhouse gases other than CO_2) and that improve environmental standards beyond those of the *Acquis* (or for sectors or plants for which a transition arrangement has been agreed, see Section 3.1). Examples of such activities can be found in:

– renewable energy sources,
– energy efficiency in a built environment,
– transport,
– combined heat and power, and
– industrial waste management.

Despite the fact that these activities are not subject to double counting risks (or at least to a much lower risk), their GHG accounting procedures are generally considered to be relatively complex.[22] The main element of these procedures is the determination of a reference scenario (a baseline), which shows what the emissions would have looked like in the absence of the project. The complexity of a baseline is that it describes a hypothetical scenario with the most reasonable alternative to the project; the scenario is hypothetical because it will be replaced by the project. Carrying out a JI project in, e.g., the transport sector or a built environment would in principle imply that a baseline would be needed for each lorry or car or building. Moreover, it would imply a very complex system of monitoring the project's performance with high transaction costs.

One way to avoid these transaction costs is to standardize procedures so that they could be applied to a multitude of projects in the same category. From an

---

[22] Van der Gaast, 2005.

investment point of view, standardization is considered desirable because it increases transparency. Ideally, with standardized baselines, each investor in a JI project in a particular host country would be able to use the same baseline for the same types of projects. However, the determination of a standardized/multi-project baseline in itself may require more time and data than one project-specific baseline, as it generally requires a macro and/or sector analysis.

Under JI Track I (see for a difference between JI Track I and Track II, box 4.1), the application of standardized multi-project baseline and monitoring procedures is possible as long as both the host and investor countries agree on their use. The JISC does not play any role in this respect, and countries are free to decide how the project validation and verification is arranged. Project developers that are required to follow Track II could also apply standardized procedures, but the methodologies for multi-project baselines and monitoring must be approved by JISC before they can be used.[23]

From actual practice, several lessons can be learned with respect to baseline standardization. First, the guidelines document for the Dutch ERUPT programme contains multi-project baseline coefficients for power sector projects in a number of Central and Eastern European countries. Second, under the CDM, several projects have been proposed in the category of grid-connected renewable energy activities. In most cases, the methodologies proposed for these projects aim at determining baseline emission factors for the grid to which the project intends to deliver. These methodologies, which generally aim at identifying which of the existing and/or planned grid-connected capacity would be the first to be replaced when new capacity becomes available (i.e., the marginal capacity), strongly resemble multi-project methodologies. Finally, the CDM Executive Board has taken the initiative to consolidate similar methodologies into aggregated baseline methodologies for particular project categories, which is a form of standardization in itself.

Below are a few examples of ways to standardize baselines for JI projects and how these relate to the EU policies mentioned in Section 3. Some examples refer to project activities in Bulgaria, based on an earlier analysis of what a multi-project baseline for JI projects in this country would look like.[24]

Box 4.1. JI Track I vs Track II procedures

When both countries in a JI transaction comply with the eligibility criteria listed in the *Marrakech Accords*,[25] they can bilaterally, so without the approval of the JISC, determine a baseline for the project and deal with the project monitoring. This procedure is called Track I and in order to become eligible for this track, both Parties involved in the project must:

---

[23] It must be noted that the use of Track I is voluntary and that some countries may still decide, although they qualify for Track I, to use Track II. Possible reasons for this could be that countries may prefer third-party validation and verification as this might make them feel safer with a view to possible overselling of ERUs under Track I.

[24] See Van der Gaast, 2005.

[25] UNFCCC, 2002.

- be a Party to the Kyoto Protocol (i.e., through ratification).
- have calculated their assigned amount of GHGs (i.e., Mtonnes CO_2-eq. in 1990).
- have established a national registry for tracking the assigned amount units.
- have a national system in place for estimating actual greenhouse emissions and/or removals
- have submitted to the UNFCCC Secretariat their most recent inventory of GHG emissions, and
- have an accurate accounting of assigned amount.

Countries not in compliance with these requirements (although all countries must comply with the first three in any case) must have their projects validated by independent entities accredited by the JISC. This procedure is called JI Track II.

Track I projects are considered simpler as their project cycle does not include external validation of the project plan and verification of the project performance. This also implies that fewer transaction costs need to be made. The reason for allowing Track I projects without external verification of the project results is that the six requirements above guarantee comprehensive and reliable systems for registration of the greenhouse gas emissions in the countries involved and of the ERU transfers. Consequently, both the emission reduction from the project and the resulting transfer of ERUs are clearly registered by the countries. Any miscalculation at the project level, e.g., the ERUs transferred to the investor country are less than the actual emission reductions by the project, will thus be registered by the system. In whatever way this mistake is corrected, the key point is that it is registered. Countries that cannot properly register their emission reductions will not be able to discover such mistakes. Through independent, third party project validation and verification this risk can be largely reduced (although due to the hypothetical character of the baseline, the risk can never be fully removed).

4.2   Grid-connected Renewable Energy Sources projects

This category refers to electricity production plants built on new sites and that deliver electricity to the grid. By doing so, the project replaces electricity capacity that would otherwise have been connected to the grid. For the determination of the baseline, it is important to identify how the replaced electricity would otherwise have been produced. As the physical capacity replaced cannot easily be identified, if at all, the baseline contains an aggregated emissions factor for the entire grid. What this factor will look like depends on the country context. For instance, a JI project delivering electricity to a grid which already has an overcapacity is likely to replace already existing capacity, whereas a project delivering to a grid operating at full capacity is likely to replace planned new capacity. In this respect, JI projects could learn from the CDM where many renewable energy projects have been implemented on new sites with delivery to the grid.

The Dutch ERUPT programme provided multi-project baseline emission factors for electricity production projects to be carried out in Central and Eastern Europe. An illustrative example of the application of these factors and how they had to be adjusted following the *Acquis Communautaire* negotiations is provided below for projects in Bulgaria.

The ERUPT baseline emission factors for the Bulgarian power sector were calculated in 2001, using best available data (mostly from the end of the 1990s). As a start, ERUPT considered the then existing Bulgarian energy mix, from which it removed so-called must-run technologies (with low operational costs), such as hydro or nuclear energy, as these are unlikely to be dispatched when new capacity becomes available. In the ERUPT method, variable costs are an important determinant for dispatch, which implies that higher cost technologies such as oil and gas are assumed to be dispatched from the system before coal-based technologies. Finally, when calculating the ERUPT factors, no precise information was available on the EU pre-accession process in Bulgaria. Therefore, ERUPT assumed a linear trend from actual standards to EU best available techniques. This resulted in a baseline for power sector projects in Bulgaria of 814 gCO_2/kWh in 2005 and 689 gCO_2/ kWh in 2012.

In the meantime, Bulgaria proceeded with the pre-Accession process, leading to EU accession in January 2007. For the power sector, an important implication of this process is that four nuclear units in Bulgaria will have to be terminated. According to the Bulgarian National Communication to the UNFCCC (2002),[26] these plants will be replaced with lignite/coal and natural gas .combined technologies. This development leads to an increase in the carbon intensity of the power grid and thus adjusted baseline emission factors for Bulgarian power sector JI projects, which are much higher than the ERUPT factors: 1.19 gCO_2/kWh in 2005 and 1.08 gCO_2/ kWh in 2012 (see Table 4.1 for an overview).[27]

Table 4.1. ERUPT & MoEW calculations of the Multi-project Baseline for Power Sector JI projects, Bulgaria

ERUPT	DEPA/DANCEE – MoEW Bulgaria
• Start from existing energy mix for power	• Concrete expansion plan for Bulgaria
• Linear adjustment towards BAT	• Early termination of 4 nuclear units - replaced with lignite/coal, natural gas combined cycle
	• Planned JI projects in the baseline
• No must-run technologies in the baseline	• No must-run and peak load in the baseline
• Variable costs determine dispatch: Gas/oil before coal	• New capacity replaces off-peak power plants
• Baseline emissions factor electricity generation:	• Baseline emissions factor electricity generation:
• **2005: 814 gCO2/kWh**	• **2005: 1,190 gCO2/kWh**
• **2012: 689 gCO2/kWh**	• **2012: 1,080 gCO2/kWh**

---

[26] See <http://unfccc.int/parties_and_observers/parties/items/2249.php>.

[27] Van der Gaast, 2005.

4.3 Transport

Possible projects in the transport sector are:

- Modal shift: shifting transportation modes from, e.g., lorries to canals, commuting by car to public transport, etc.
- Fuel substitution: e.g., from diesel to compressed natural gas.
- Fuel efficiency: e.g., filters for the reduction of particulate emissions.
- Driver behavior and vehicle maintenance: e.g., through training programmes.

When determining a multi-project baseline for transport sector JI projects, a clear vision is needed of future transport sector developments in the host country. For instance:

- It is widely assumed that small operators under business-as-usual circumstances have no incentive to adjust their operations, as they generally face tight margins and need to keep afloat on a daily basis.
- The host country may have a plan to introduce road haulage charges or set vehicle standards, which would need to be included in the baseline.
- The EU standards will need to be incorporated in those JI host countries that have acceded to the EU or will do so at short notice, which depending on the timetables agreed with the European Commission will become part of the baselines.

With respect to applying JI in the transport sector, a sector-wide GHG accounting approach is more suitable than defining baseline and monitoring methodologies for each individual vehicle or company fleet.[28] Therefore, JI accounting would focus best on the estimated abatement effect of such policies as green transport plans, fuel economy programmes, vehicle emission standards, smart growth programmes, etc. A baseline for the transportation sector could be expressed as the average annual fuel consumption for vehicle types, e.g., lorries. For such calculations, current IPCC methodologies used for the *UNFCCC National Communications* could be used. In a more advanced manner, spreadsheet models can be used to estimate how employees of a particular organization go to work, e.g., for organizations with approximately 1000 employees within a particular region. As such, the spreadsheet would deliver a baseline emissions factor in terms of CO_2 per passenger per km. In the UK, such a spreadsheet model has been developed by Napier University.[29]

4.4 Waste management

In the context of JI in Bulgaria a waste management project has been constructed, which uses wood waste as a fuel for electricity in the Paper Factory Stambolijski.

---

[28] Van der Gaast, 2005; Bossi and Ellis, 2005.
[29] See Begg, et al., 2002.

The waste is from the paper production. The project is presently under validation and the Netherlands Government will purchase its ERUs. In the baseline, it is assumed that under business-as-usual circumstances, landfilling of wood waste in existing landfills would be allowed at least up to 2013. The reason for this assumption is that the waste from the factory is presently classified as non-hazardous by the Bulgarian regulations (Environmental Protection Law 2002; Waste Management Law 2003). Although these regulations prescribe that methane emitted from landfills must be captured and combusted/flared, this only applies to new landfills, whereas for existing regulations deadlines of compliance are unclear. Moreover, according to the EU Waste Landfill Directive, the activities foreseen under the project would have to be carried out anyway.[30] However, since Bulgaria has agreed with the European Commission on a transitional arrangement, this Directive will not be implemented in the country before 2012.

Consequently, the baseline for the electricity from the waste component refers to the power that would otherwise have been purchased from the grid. For this component the ERUPT multi-project baseline emission factors have been used (see Section 4.2). The baseline component for the methane reduction part of the project is more complex, but has been standardized with the help of a spreadsheet model with default values from literature, IPCC and CDM experience. This model takes into consideration the general issues associated with measuring methane capture from landfills, such as the waste composition, oxidation of gas, etc., which often differ between landfills. Other examples of ways to standardize the measurement of methane emissions from landfills are spreadsheet models such as Summerleaze and GasSim in the UK.[31]

5.   CONCLUSION

The scope for JI has been reduced since several potential JI projects in the new member states have become compulsory or creditable under EC law. On the other hand, the demand for JI projects may increase through the Linking Directive which allows EU installations to purchase ERUs for reasons of compliance with their EU ETS commitments. The extent to which this demand will be executed depends on the development of the CO_2 allowance prices within the ETS. Should EU ETS prices reach the levels during 2008-2012 as presently expected on the forward market (December 2008 allowance contracts), i.e., close to € 20 per tonne CO_2-eq., and JI credits keep floating between € 5 and 10 per tonne, JI could offer arbitrage opportunities for EU installations.

Given this political context, JI could be applied on a broad scale in Eastern European countries that are included in Annex I but which have not acceded to the EU (e.g., Ukraine, the Russian Federation, Belarus); in the new member states, JI

---

[30]   Council Directive 1999/31/EC of 26 April 1999 on the landfill of waste.
[31]   See Begg, et al., 2002.

can be applied as long as the projects are compatible with EC law. In this respect, JI could be an important complementary policy instrument for stimulating sustainable energy technologies for sectors and installations not covered by the EU ETS or that are subject to transitional arrangements in the context of the *Acquis Communautaire*. In order to effectuate this potential it is important that accounting procedures for GHG emission reductions are standardized so that methodologies can be applied to a multitude of projects and project-specific transaction costs could be reduced. For this, useful lessons could be learnt from the CDM practice.

References

Begg, K.G., T. Jackson, D.van der Horst, W. van der Gaast, J. Bandsma, M. ten Hoopen and S. Sorrell, 2002. *Guidance for UK Emissions Trading Projects, Advice to Policy Makers*, A report for the Department of Trade & Industry, University of Surrey, Guildford, UK.

Bossi, M. and J. Ellis, 2005. *Exploring Options for 'Sectoral Crediting Mechanisms'*, IEA/OECD, 2005.

Fenhann, J., 2007. *CDM/JI Pipeline Overview*, UNEP Risø, Denmark, March 2007, http: <www.cd4cdm.org/Publications/CDMpipeline.xls>.

JIN, 2000. 'The Netherlands and Romania Agree on ERU Transfer', *Joint Implementation Quarterly*, Vol. 6, No. 4, p. 3.

JIN, 2001. 'Capacity building for JI in Bulgaria and Romania for ERU-PT', *Joint Implementation Quarterly*, Vol. 7, No. 1, p. 2.

JIN, 2003. 'European Commission Proposes Linking JI and CDM credits to ETS', *Joint Implementation Quarterly*, Vol. 9, No. 3, p. 2.

Nondek, L., M. Malý, V. Splitek, and J. Pretel, 2001. *Joint Implementation in the Context of the EU Accession: the case of the Czech Republic*, PCF*plus* Report 7, Washington, D.C, USA.

Schrijver, N.J., 1995. 'Joint implementation from an International Law Perspective', in: Jepma, C.J. (ed.), 1995. *The Feasibility of Joint Implementation*, Kluwer Academic Publishers, pp. 133-134.

UNFCCC, 2002. Report of the Conference of the Parties on its Seventh Session, held at Marrakesh from 29 October to 10 November 2001, FCCC/CP/2001/13/Add.1, 21 January 2002.

Van der Gaast, W.P., 2005. 'Baseline standardisation in JI Track-I and Green Investment Schemes', paper presented at JI Track-I workshop on 7-8 September 2005, Prague, Czech Republic.

THE MARKET POTENTIAL OF LARGE-SCALE NON-CO$_2$ CDM PROJECTS

Axel Michaelowa*, Jorund Buen, Arne Eik and Elisabeth Lokshall**

ABSTRACT

The Kyoto Protocol's Clean Development Mechanism (CDM) allows industrialized countries with emission reduction commitments to generate emission credits though greenhouse gas reduction projects in developing countries. CDM has so far commonly been regarded as a mechanism for CO$_2$ reduction projects; however, the bulk of CDM credits could come from non-CO$_2$ reduction projects. The reason is that such projects are large and hence have low transaction costs in relative terms, emission reduction costs are low, and determining project 'additionality' is simple since reducing non-CO$_2$ gases normally does not entail any direct benefits. The reduction of HFC-23 from the production of refrigerant HCFC-22 and the destruction of N$_2$O from adipic acid production will generate hundreds of millions of emission credits by 2012. Projects capturing methane from landfills and coalmines or reducing gas flaring from oil production and N$_2$O from nitric acid production yield smaller, but still substantial reductions compared to 'classical' renewable energy projects. We assess the technical potential until 2012 for these CDM project types and discuss potential implementation barriers. Based on a detailed case study of HFC-23, we suggest possible implications for the other project types. Interest groups promoting other project types can stall the development of non-CO$_2$ projects; this is already occurring concerning HFC-23 reduction projects. China, having a sizeable potential in all non-CO$_2$ project types covered, has introduced a tax for such projects due to their low reduction costs and limited local sustainable development benefits.

1. INTRODUCTION

The Clean Development Mechanism (CDM) under the Kyoto Protocol allows industrialized countries with emission targets to generate emission credits (Certified Emission Reductions, CERs) through emission reduction projects in developing countries. Initial estimates of developing country potential to supply large amounts of CERs within a short time were pessimistic. Most analysts thought that renewable energy and energy efficiency projects would generate the bulk of CERs (see, e.g., Jotzo and Michaelowa 2002).

* Institute of Political Science, University of Zurich, Switzerland.
** Point Carbon, Oslo, Norway.

W.Th. Douma et al., eds., The Kyoto Protocol and Beyond
© *2007, T·M·C·ASSER PRESS, The Hague, The Netherlands and the Authors*

After a slow start due to a lack of CER demand from industrialized countries, an increasing number of governments have set up CER acquisition programmes. Moreover, private companies in Europe can use CERs to fulfill their obligations under the EU emissions trading system. Japanese companies have been eager to buy CERs to hedge against future policy requirements. Overall, in early 2007 over € 4.5 billion had been committed or already spent on CER acquisition. Likewise, the supply of CERs has exploded. Over 600 projects have been registered by the CDM Executive Board, 1400 projects have been submitted to accredited operational entities for validation and over 200 baseline methodologies have been proposed for a wide range of technologies.

Surprisingly, over the last three years, several non-CO_2 reduction options have emerged that will dwarf all CO_2-related projects, at least in the short and medium term. These technologies can be installed quickly, have low costs and no risk of failing the additionality test as most of them are not linked to a production of goods or services. Despite the surprise, surveys on the potential of non-CO_2 gas reduction already forecast large volumes in the past. IEA GHG R&D (2003) estimated a sizeable potential for several technologies; about 1.1 billion t CO_2 eq. per year if additionality is not checked and about half of that volume when there is a rigid additionality check (see Table 1).

Table 1. Overall reduction potential in Non-Annex I countries for different technologies

Project type	Reductions at negative costs, i.e. with additionality problems (million t CO_2 eq.)	Reduction volume at a price of $ 10 (million t CO_2 eq.)	Cost range per CER ($)
Wastewater methane capture	318	318	<0
Landfill gas capture	91	266	< 0 -5
Coal-bed and mine methane	81	221	<0 -4.5
Pipeline leakage reduction	86	150	Wide range
N_2O reduction from nitric acid	0	107	0.5-1
Gas flaring reduction	0	75	4-6
N_2O reduction from adipic acid	0	23	0.2
HFC-23 reduction from HCFC-22	0	14	0.3
SF_6 reduction	0	10	0.7-3
PFC reduction	7	7	Wide range
Sum	**583**	**1191**	

Source: Marginal abatement cost curve levels for 0 and 10 $ from IEA GHG R&D (2003)

So far, wastewater has only been addressed in the CDM context in small-scale projects. Likewise SF_6 reduction from electric equipment appliances would only generate small-scale projects and thus these categories are not discussed in this article. We will assess the current status of each of the other technologies in the CDM context according to project size, their costs per t CO_2 equivalent and the potential for their future expansion.

2. HFC-23 DESTRUCTION

HFC-23 (trifluoromethane, CHF_3), a greenhouse gas with a GWP of 11,700, is generated as a by-product during the manufacture of HCFC-22 (chlorodifluoromethane, $CHClF_2$). The quantity of HFC-23 produced depends in part on how the process is operated and the degree of process optimization that has been performed. At plants not fully optimized, the upper band for HFC-23 emissions is 3 to 4% of HCFC-22 production. However, many plants have implemented process changes in recent years to reduce HFC-23 generation because the generation of more HFC-23 means that a less valuable product (HCFC-22) is produced. At these plants, the likely range of emissions is about 1.5 to 3% of production with 2% being a reasonable average estimate (IPCC 2000). The IPCC default value of 4% may thus considerably overestimate HFC-23 production.

There exist four different proven continuous process technologies for the decomposition of HFCs. The investment costs of the Solvay process are € 3 million to destroy 200 metric tonnes of HFC-23 per year plus € 0.2 million annual operating costs. On this basis, Harnisch and Hendricks (2000) report abatement costs of 0.2 €/t CO_2 eq. The gaseous/fume oxidation process used for the Ulsan CDM project in South Korea (see below) quotes destruction costs of 2.5-4 €/kg HFC which would mean abatement costs of 0.2 – 0.4 €/t. Costs of the PLASCON process reach 2.5 – 3.5 €/kg HFC, and for the Ohei process 1.5 –2.5 €/kg HFC (UNEP TEAP 2002). The cost range of 0.2 – 0.5 € / t CO_2 eq. is thus robust and also confirmed by Jimenez (2005). Lead times for the installation of these technologies amount to 12 months; the lead time from planning a CDM project to obtaining CERs is likely to be 18 to 24 months.

The first approved baseline methodology of the entire CDM process relates to HFC destruction. It was submitted by the Ulsan project developers and approved in July 2003 under the formal title AM0001 'Incineration of HFC-23 Waste Streams'. For about one year nobody found a problem with HFC-23 reduction and the Ulsan project developers set up their destruction plant which was inaugurated in May 2004. Suddenly experts on ozone depleting substances raised the alarm (see Schwank 2004) and triggered an unprecedented process of baseline revision. This was due to the fact that the Chinese government had organized a workshop on HFC-23 reduction in February 2004 that for the first time drew the attention of a wider audience to the huge CER potential from that technology.

The main reason for alarm was that HCFC-22 is a first-order replacement for the 'ozone killers' CFC but remains an ozone-depleting substance. Thus it has been phased out in industrialized countries but developing countries are allowed to increase production until 2015 and to continue production at the 2015 level until 2040. Schwank (2004) argues that CER sales allow HCFC-22 producers to expand production indefinitely as the CER revenues alone make HCFC-22 production profitable even if the HCFC-22 is given away for free (or vented). This problem could be solved by requesting proof that the HCFC-22 is actually sold and used (Perspectives 2004). Moreover, Jimenez (2005) argues that the Montreal Protocol does not account for the use of HCFC-22 as a feedstock (where most of the growth will come from). As the export of products containing HCFC-22 to world markets is not allowed, a reduction in costs would not result in a substantial increase of Montreal-Protocol-related sales and production up to 2015. Another serious problem is that HCFC-22 is a greenhouse gas with a GWP of 1300 which is not covered under the Kyoto Protocol. Moreover, arguments were made that show a misunderstanding of the function of markets. The indirect pressure on other competing companies world-wide under pressure to come forward with similar CDM project proposals as soon as possible is seen as negative while it is an ideal way for the market to promote an incentive for emission reductions.

In parallel, HCFC-22 producer DuPont made a non-public submission to the CDM Executive Board (EB) arguing that the baseline default HFC-23 production rate should not be set at 4% of HCFC-22 production but at just the 1.37% achieved at DuPont's Louisville Works (DuPont 2004). The reason for DuPont arguing for such a strict baseline emission factor was that it feared that its competitive position would deteriorate due to the fact that it could not implement any HFC-23 reduction CDM projects while its competitors could do so. Jacob (2005) argues that DuPont feared that HFC-23 reduction could be made ineligible, that it was helpful in avoid-ing such an outcome and that a large programme of projects in China could still lead to such a decision.

In July 2004, the Swiss EB member launched a review request for AM 0001 which was put on hold without further notice. This decision without prior consulta-tion raised a lot of protest and thus the EB belatedly launched a public consultation in September 2004. A total of 22 submissions were made, many of which com-plained about the process. In December 2004, the EB decided that AM 0001 should apply only to HCFC-22 production facilities with at least three years of operating history by the end of the year 2004 and that baseline HCFC production would be capped at the maximum (annual) production during the last three years. The HFC-23/HCFC-22 rate would be capped at 3%. If no direct measurement of HFC-23 release or mass balance exists for these three years, the default rate would be set at 1.5, i.e., almost the rate proposed by DuPont. The EB asked the COP to take a decision on capacity expansion projects.

UNEP TEAP (2003) estimates HCFC-22 production in Non-Annex B countries at 166,000 t in 2002 and projects it to rise to 181,000 t in 2005 and 205,000 t in 2010. Unfortunately, there are no reliable production data on the plant level, al-though it is known that one plant operates in both Argentina and Mexico, three plants in India and between 11 and 19 in China (UNEP 2001, Chinese Chemical Network 2004, company websites, participants' lists of Chinese HFC workshop January 2005, Quimobasicos 2005). The last publicly available complete survey dates from 1995. Recent data from China suggest a much faster rise in production than expected so far (CAOSMI 2003, Lu 2004, Clini 2005), reaching between 178 and 200,000 t at the end of 2003. At a 2.9% HFC-23 generation rate, the 2003 abatement level would reach around 80 million CERs, reaching over 110 million in 2010. Assuming that data on the HFC-23/HCFC-22 rate do not exist and thus the rate of 1.5% has to be used, a conservative estimate of the CER potential in HFC-23 destruction amounts to about 40 million CERs per annum which can be harnessed until 2008 at costs of less than 0.5 €/CER. Calviou (2005) argues that historical data are generally available – then the potential would be around 75 million CERs p.a. Additionality, testing is simple as there is no relevant market for HFC-23 and the HFC-23 destruction entails costs that would not be incurred otherwise. Monitoring is relatively complex but manageable for operators of the plant.

By May 2005, the Ulsan and another project in India had formally been registered by the CDM, estimating 39.8 million CERs until 2012. However, the first monitoring report on Ulsan in April 2005 only claimed 0.9 million CERs, 30% of the estimate for the period. The Italian government which already signed a MoU with the Chinese State Environmental Protection Agency in October 2003 and the World Bank were eager to develop all Chinese HFC-23 projects together. However, the Chinese government was not willing to sell them all the CERs. In late 2005, China decided to tax CER revenues from HFC-23 projects at a rate of 65%. Nevertheless, the submission of projects went on smoothly. By 1 May 2007, 15 HFC-23 projects with a cumulative CER volume of 382 million by 2012 had been registered and 5 projects with a cumulative volume of 123 million CERs were under validation. Annual CER estimates were 68 and 22 million, respectively.

3. DESTRUCTION OF N$_2$O FROM ADIPIC ACID PRODUCTION

Adipic acid is used as a feedstock for the production of polymers, especially nylon. Through relatively simple catalysts or combustion, N$_2$O emissions can be virtually eliminated. The reaction is strongly exothermic and the heat produced must thus be removed; if there is a suitable demand on the production site, then it may be recovered and used to produce steam (Hendricks and Bode 2000). The catalytic destruction achieves a destruction of 90-95%, the combustion 98-99% (IPCC 2000). AEAT (1998) estimates reduction costs of the catalytic route at only 0.1 €/t CO$_2$ eq.

There are six adipic acid plants in developing countries (one each in Brazil, Korea and Singapore, three in China) with a 2003 production capacity of 456,000 t adipic acid, which expanded to 659,000 in 2005 (Chemical Week 2003, 2005). The catalytic process used by DuPont in Canada leads to a reduction of 87 t CO$_2$ eq./t adipic acid produced at an efficiency of 97% (ICF 2002). The gross emissions factors of the DuPont plants in Texas and Singapore, which have abatement technologies installed, reach 98 t CO$_2$ eq./t adipic acid (DuPont 2001). IPCC (2000, p. 3.33) gives a gross emissions factor of 93 t CO$_2$ eq. which then has to be multiplied by the efficiency factor of the abatement technology times its utilization factor to derive the emission reduction. Applying the factor 87 to be conservative, CER potential reaches 30 million for 2003 production and 47 million for 2005 levels. Additionality testing is not a problem as there is no market for N$_2$O of less than perfect purity and the destruction entails costs that would not be incurred otherwise.

The French company Rhodia submitted a baseline methodology for this project type in June 2004 which was approved surprisingly quickly in February 2005 as AM 0021. The methodology submission had incorporated lessons from the debacle of AM 0001 and thus defined the methodology as applying only to existing capacity, given that Rhodia's plants do not envisage a capacity expansion. However, a methodology for capacity expansion should not face the same problems as in the case of HFC-23 as adipic acid is neither an ozone-depleting substance nor a greenhouse gas.

By 1 May 2007, 2 projects with a cumulative CER volume of 92 million by 2012 had been registered, whereas 2 projects with a cumulative CER volume of 77 million were under validation. Annual CER estimates were 15 and 14 million, respectively.

4. DESTRUCTION OF N_2O FROM NITRIC ACID PRODUCTION

Nitric acid plants produce N_2O as a by-product of the high temperature catalytic oxidation of ammonia (NH_3). The N_2O emission factor depends on the pressure under which the nitric acid is produced and can vary from 4.2 kg N_2O/t nitric acid (1.3 t CO_2 eq.) at low pressure to 9.7 kg N_2O/t nitric acid (3 t CO_2 eq.) at high pressure. The N_2O can be reduced by approximately 80% utilizing a catalyst which is produced by BASF and has successfully been used in three facilities since 1997 (secondary reduction). The N_2O can also be burned after its generation (tertiary reduction).

Unfortunately, disaggregated nitric acid production data are not publicly available. For the production of developing countries' emissions estimates range from 15 to 34 million t CO_2 equivalent (OECD 2003, USEPA 2005). Egypt is the only country with plant-specific data; the CER potential of four plants is estimated at 5 million per year (Samir 2005). As in the case of N_2O abatement from adipic acid production, the additionality test is simple. The only challenge may be to define the baseline emissions factor as the variability of N_2O production seems to reach 50% and thus a lengthy pre-project monitoring may be required.

Astonishingly, it took a long time for this project category to be developed under the CDM. A company called CarbonVentures claimed that it had an exclusive agreement with BASF to develop and implement N_2O abatement projects at fertiliser and chemical plants worldwide. However, BASF's technology competitor Uhde managed to collaborate with the Austrian consultancy Carbon to submit the first methodology proposal in April 2005 linked to a PDD addressing the largest nitric acid plant in the world located in Abuqir, Egypt with 1.4 million CERs p.a. The methodology was approved as AM 0028 in February 2006. In November 2005, the German consultancy N.Serve submitted a methodology for a tertiary reduction and had it approved in July 2007 as AM 0034.

By 1 May 2007, 4 projects with a cumulative CER volume of 24 million by 2012 had been registered (all under AM 0028), whereas 23 projects with a cumulative

CER volume of 44 million were under validation (7 under AM 0028 and 16 under AM 0034). Annual CER estimates were 4 and 8 million, respectively.

5. LANDFILL GAS CAPTURE AND DESTRUCTION

The amount of landfill gas (LFG) generated in a landfill is not only linked to the quantity of waste but strongly depends on its composition. It is thus very situation-specific and varies strongly over time. For a relatively wet, deep but not well managed landfill in a middle-income developing country, already operating for ten years and receiving a million t of waste annually, emissions from the last year's waste amount to 140 kt CO$_2$ eq. whereas the first year's waste still emits 52 kt CO$_2$ eq., summing up to 0.9 million t CO$_2$ eq. Collection efficiency is usually about 80%; i.e., 0.7 million CERs could be achieved. Lead times for setting up LFG collection equipment are about one year (Redding Energy Management 1999). If planned in tandem with the design of a new landfill, they do not involve any relevant delay.

Low-income countries normally produce between 0.15 and 0.25 t municipal solid waste per capita per year, countries beyond 1000 € GDP/capita between 0.25 and 0.4 t (World Bank 1999). For a metropolis of 10 million people, this translates into 1.5-2.5 million t of waste for low-income countries and 2.5-4 million t for middle-income countries. Centralized collection of the waste in one or two landfills is a precondition for a large LFG collection project. Based on 80 ha/1 million t and a project lifetime of 15 years, costs would amount to 0.4 €/CER. Summing up the 27 metropolises in developing countries with more than 1 million t of waste per year and excluding those that already collect landfill gas, an overall annual potential of 44 million CERs is estimated.

In comparison to the other project categories, landfill gas projects face a number of barriers. First, municipal waste management in developing countries is usually a highly politicized issue and often prone to corruption. Often, many different agencies and political levels are involved, leading to long times to negotiate changes. Moreover, the potential is squeezed from two sides: the analysis of the different cities shows that in newly industrialized countries (Hong Kong, Korea) LFG recovery and burning is already done. Moreover, incineration is increasingly used which eliminates methane generation of landfills. In lower income countries, the management of the landfills is often erratic, corruption endemic and it is thus difficult to achieve a reliable methane generation (Manila, Jakarta). The challenge for CDM project developers is thus to find developed projects in the intermediate bracket. In the upper middle-income countries (Brazil) incineration starts to be used but the landfills are relatively well managed and thus a good target for methane recovery. There are three baseline methodologies available for LFG collection. They have been favored by the ease of additionality testing. Moreover, there is a consolidated methodology (ACM 0001).

By 1 May 2007, 45 projects with a cumulative CER volume of 85 million by 2012 had been registered, whereas 86 projects with a cumulative CER volume of

96 million were under validation. However, the first LFG projects asking for the issuance of CERs have had a much lower issuance than projected in the project documentation (10-15%).

6.   COALMINE AND COAL-BED METHANE CAPTURE/DESTRUCTION

Underground coalmines generate methane that seeps from coal seams and is a serious safety hazard. For this reason, shafts have to be ventilated to keep methane concentration below a certain level in order to prevent explosions. Due to the low concentration, the resulting coalmine methane (CMM) is usually vented into the atmosphere. Only recently, coalmines in industrialized countries have started to use the methane for small cogeneration plants (1-10 MW). The inclusion of coalmine methane in the renewable energy act in Germany led to a mushrooming of such plants; between 1999 and 2003 90 MW were installed in the former mining districts. Electricity generation costs are around 6 ct/kWh (Backhaus, et al. 2003). The term 'coalbed methane' (CBM) is only used for activities where unmineable coal-seams are drilled to actively extract the methane.

All developing countries with a large underground mining industry have a relevant potential for coalmine methane reduction. In 2000, worldwide CMM emissions totaled 440 million t CO_2 eq., expected to increase to 561 by 2020, of which 176 are in China, 26 in North Korea and 9 each in India and South Africa (Methane to Markets Partnership, 2005). China has been the focus of US activities since the mid-1990s to open up a market for US coalmine methane reduction technologies. In 1995, Jinsheng Anthracite Mining Company used seven CMM wells to fuel a 400kW power plant. The company increased the capacity by 22 MW in early 2002 by building a second CMM power plant. Jinsheng wants to expand the plant to 120 MW, the largest CMM CDM project to date generating 1.6 million CERs annually. In 2005, there were five Chinese projects aiming for CDM status, each of which had submitted its own new methodology. The EB thus asked a consultant to analyze these methodologies and to provide recommendations for consolidation, which was done as ACM 0008.

By 1 May 2007, 3 projects with a cumulative CER volume of 9 million by 2012 had been registered, whereas 28 projects with a cumulative CER volume of 119 million were under validation. All these projects but one are located in China. India which also has a large coalmining industry has so far not been able to mobilize coalmine methane project developers due to a highly complex regulatory structure.

7.   GAS FLARING REDUCTION FROM OIL PRODUCTION

Under present oil production practices in many developing countries, associated gas is still seen as a waste by-product of oil which can hinder the oil flow. In Sub-Saharan Africa, it is generally disposed of through flaring or venting for lack of adequate markets or a lack of an institutional and regulatory framework to support its utilization. The issue in developing associated gas recovery is purely economic, not technical, as associated gas can physically be separated from oil for further use within or outside the industry. Own use includes re- injection, either to improve oil flow, the reservoir productivity and thus increase its useful life, or to be kept in storage for later use. External usage directs gas toward consumption markets, either domestic or international, where it can be used as a fuel or feedstock (UNDP and World Bank 2001).

Reducing gas flaring to the average level of industrialized countries could generate 105 million CERs annually in Sub-Saharan Africa, with Nigeria accounting for three quarters. The Middle East would get 57 million, Latin America 42 and North Africa 21. Asia's share would be limited to 6 million. The problem with gas flaring is that its incidence is inversely proportional to the governance of a country. The highest potential can be found in chaotic countries that have not ratified the Kyoto Protocol nor set up a national CDM authority. The World Bank (2003) is supporting project development in Algeria, Indonesia and Nigeria but so far has run into institutional problems. Additionality, determination is difficult as the saved gas has an economic value.

By 1 May 2007, 2 projects with a cumulative CER volume of 16 million by 2012 had been registered, whereas 17 projects with a cumulative CER volume of 52 million were under validation.

8.   PFC EMISSIONS REDUCTION FROM ALUMINIUM PRODUCTION

So-called anode effects, unplanned interruptions to the process producing aluminium generate the high-GWP perfluorocarbons CF_4 and C_2F_6. Aluminium producers try to reduce anode effects as they reduce productivity. So in contrast to HFC-23 and N_2O reduction, additionality determination is a hurdle. There are massive differences in the emission intensities of the five main technologies employed for aluminium production, ranging from 0.8 (Point Feed Prebake) to 7.9 t CO_2 eq. (Side Worked Prebake) per t aluminium (International Aluminium Institute 2002); however, the share of the high-emission technology is only 6% and declining. Annual global CER potential amounts to 9 million, 40% of which would be located in China. Two projects in India and Argentina have been proposed; however, their methodologies were rejected. Based on this experience, only 2 million CERs per year are expected.

9.   OVERALL CER POTENTIAL UNTIL 2008 AND THE IMPACT ON THE CDM MARKET

The seven technologies covered can quickly generate huge low-cost amounts of CERs given lead times of 1.5-2 years (see Table 2). The main issue is whether projects can get quick host country approval. This will be the major hurdle as the sustainable development benefits in terms of environmental and social improvements are likely to be fairly low. For the HFC-23 and N_2O destruction technology, a high-tech process with low labor intensity and no improvement concerning local pollutants will be used. For landfill gas collection, electricity generation may be requested. Coalmine methane projects will generally improve safety and thus have high sustainability benefits.

Table 2. CER potential of large non-CO_2 reduction CDM projects

Technology	Costs €/ CER	Lead time (months)	Approved methodology	CER potential 2004-2007 (million)[1]	Annual CER potential 2008 – 2012 (million)	Number of projects
HFC-23 destruction	0.2 – 0.5	24	Yes	140	90	20
N_2O - adipic acid	0.1	24	Yes	110	50	5
LFG destruction	0.4	18	Yes	30	15	20[2]
N_2O - nitric acid	0.1	24	Yes	20	15	40
CMM	0-4	18	Yes	10	20	40
Gas flaring red.	<0-?	?	Yes	10	10	10
PFC red.	<0-high	24-36	No	<5	2	8
Sum	**<0.5**	**18-36**		**~320**	**~200**	**~140**

[1] Assumption of a full utilisation after the expected lead time, only countries above corruption index of 3.5 for landfill gas.

[2] Large projects above 1 million CERs/year

Overall, the large projects can generate 300 million CERs including 2007 and a total of 1.3 billion CERs until 2012 at total costs below 1 €/CER. Due to the concentration of the really large projects in the four host countries of China, India, Korea and Brazil, they lend themselves to cartelisation and/or direct sales to Annex B governments.

References

AEAT (1998): Options to Reduce Nitrous Oxide Emissions, AEAT-4180: Issue 3.
Backhaus, Clemens; Meyer, Jürgen; Spangardt, Gorden (2003): ,Energetische Nutzung von Grubengas. CO_2 Reduktion durch Methannutzung', Terratec, Leipzig.
Ball, Jeffrey (2003): 'Changing Climate: New Market Shows Industry Moving On Global Warming - Even as Bush Opposes Kyoto, Firms Are Trading Rights To Emit Greenhouse Gases - DuPont Tries to Get Out Front', *The Wall Street Journal*, 16 January.
CAOFSMI (China Association of Organic Fluorine and Silicone Material Industry) (2003): Survey Report on HCFC-22 production in China, Beijing.
Calviou, Louise (2005): 'HFC-23 CDM projects', presentation at Sino Italian workshop on HFC 23 CDM Projects in China, Beijing, 24 January 2005.
Chemical Week (2003): Adipic acid, 23 April, p. 25.
Chemical Week (2005): Adipic acid, 4 May, p. 31.
China CDM Federation (2005): China CDM Business Delegation, 2005 Carbon Expo, Cologne.

Chinese Chemical Network (2004): <www.chinachemnet.com>.

Clini, Corrado (2005): 'The policy and plan on CDM', presentation at Sino Italian workshop on HFC 23 CDM Projects in China, Beijing, 24 January 2005.

DuPont (2004): 'Change in HFC-23 Emission Reduction Project Baseline Assumptions (AM-0001)', Letter to CDM Executive Board, 3 June 2004.

DuPont (2001): 'Nitrous oxide abatement at DuPont's adipic acid facilities', Missisauga.

Greenpeace (1995): 'The ozone layer destroyers', Amsterdam.

Harnisch, Jochen; Hendricks, Chris (2000): 'Economic Evaluation of Emission Reductions of HFCs, PFCs and SF$_6$ in Europe', Brussels.

Hendricks, Chris; Bode, Jan-Willem (2000): 'Prioritising options to reduce non-CO$_2$ emissions through the Kyoto Mechanisms in different countries', Ecofys, Utrecht.

Hu Jianxin (2005): 'Reduction of HFC-23 as a a GHG and phaseout of HCFC-22 as a ODS', presentation at Sino Italian workshop on HFC 23 CDM Projects in China, Beijing, 24 January 2005.

ICF (2002): 'Greenhouse Gas Emission Reduction Verification Audit for DuPont Canada Inc's Maitland Ontario Apidic Acid Plant', Toronto.

IEA Greenhouse Gas R&D Programme (2003): 'Building the cost curves for the industrial sources of non-CO$_2$ greenhouse gases', Report No. PH4/25, Cheltenham.

International Aluminium Institute (2002): 'PFC emissions reduction programme 1990-2000', London.

IPCC (2000): 'Good practice guidance and uncertainty management in national greenhouse gas inventories', Geneva.

Jacob, Tom (2005): 'Baseline Methodologies for HFC-23 Projects – DuPont View', presentation at Sino Italian workshop on HFC 23 CDM Projects in China, Beijing, 24 January 2005.

Jimenez, Guillermo (2005): 'Policy alternatives for abatement of HFC-23 emissions in China', presentation at Sino Italian workshop on HFC 23 CDM Projects in China, Beijing, 24 January 2005.

Jotzo, Frank; Michaelowa, Axel (2002): 'Estimating the CDM market under the Marrakech Accords', in: *Climate Policy*, 2, p. 179-196.

Lu Guoqiang (2004): 'The prioritized CDM projects in China', in: DEG, TÜV Rheinland (eds): Presentations Sino-German Seminar on 'CDM projects: opportunities and Financing', May 17, Chengdu, p. 44-48.

Marks, Jerry (2003): 'The Aluminium Sector Story', <http://unfccc.int/files/meetings/archive/application/vnd.ms-powerpoint/wbcsd_11_05.ppt>.

Methane to Markets Partnerships (2005): 'International opportunities for project development: recovery and use of methane from coal mines', Washington.

OECD (2003): 'IEA CO$_2$ emissions from fossil fuel combustion', Paris.

Perspectives Climate Change (2004): 'Comment on the baseline methodology for HFC 23 reduction from HCFC 22 production (AM 0001)', Hamburg.

Quimobasicos (2005): 'Quimobasicos HFC recovery and decomposition project', Monterey.

Redding Energy Management (1999): '2% Renewables Target in Power Supplies', Canberra.

Sami, Tarek (2005): 'Technology Profile & Market Analysis for Reduction of Nitrous Oxide at Egyptian Fertilizers Companies (Survey Results)', Presentation at CD4CDM Workshop for Fertilizers Sector in Egypt, 5-6/4/2005, RAMSES Hilton, Cairo, <http://www.cdm egypt.org/Fertilizer.htm>.

Schwank, Othmar (2004): 'Concerns about CDM projects based on decomposition of HFC-23 from 22 HCFC production sites', Infras, Zurich.

UNDP and World Bank (2001): 'Africa Gas Initiative. Main Report', Washington.

UNEP (2001): 'Inventory of Trade Names of Chemical Products Containing Ozone Depleting Substances and their Alternatives', Paris.

UNEP Technology and Economic Assessment Panel (2002): 'Report of the TEAP Report of
the Task Force on Destruction Technologies', Nairobi.

US EPA (2005): 'International Non-CO$_2$ Greenhouse Gas Marginal Abatement Report',
Washington.

World Bank (2003): 'Global Gas Flaring Reduction. Kyoto Mechanisms for Flaring Reductions', Report No. 2, Washington.

World Bank (1999): 'What a waste: solid waste management in Asia', Urban Development
Sector Unit, Washington.

LEGAL NATURE OF KYOTO UNITS

Matthieu Wemaëre*

The Kyoto Protocol is based on the principle of achieving quantitative reductions in greenhouse gas (GHG) emissions. Article 3, paragraph 1 establishes binding emission reduction or limitation targets for the period 2008-2012 on so-called 'Annex I Parties', which are listed in Annex B to the Protocol. The Protocol provides Annex I Parties with certain flexibility in order to achieve these emission reduction targets, through the possibility to trade emission rights (AAUs, Assigned Amount Units) pursuant to Article 17 and project-based emission credits generated through JI or CDM projects, which are eligible under Articles 6 and 12.

Contracting Parties need legal certainty for implementing the Kyoto Protocol in good faith. Both Governments and private entities should be provided with security and confidence concerning any market-based instrument so as to promote their involvement and increase market liquidity. In particular, Parties and legal entities concerned with transfers of JI/CDM credits need to know what is acquired/sold through contractual arrangements (Emission Reduction Purchase Agreements), and what will be the implications (property, accounting, taxation) for their business activities.

However, the international climate regime (UNFCCC supplemented by the Kyoto Protocol and modalities adopted by the Contracting Parties for its implementation, in particular the so-called Marrakech Accords) does not define the legal nature of Kyoto Units.

Such a discussion on the legal nature of tradable units created by International Law through a multilateral environmental agreement is also relevant for tradable units created through domestic measures aimed at implementing the Kyoto Protocol, such as EU emission allowances allocated to some capped sectors through the EU Emissions Trading Scheme (hereinafter EU ETS) established by EC Directive No. 2003/87/EC, as amended by the Linking Directive No. 2004/101/CE. As EU emission allowances are AAUs with an EU tag as of 2008, those concerned with the EU ETS need to know what they are precisely allocated, and what can be transferred on the EU market.

---

* Senior Lawyer, Partner, Huglo Lepage & Partners Law Firm, Director of the Brussels Office. See also Matthieu Wemaëre and Charlotte Streck, 'Legal Ownership and Nature of Kyoto Units and EU Allowances' in *Legal Aspects of Implementing the Kyoto Protocol Mechanisms*, Oxford University Press 2005.

W.Th. Douma et al., eds., The Kyoto Protocol and Beyond
© 2007, T·M·C·ASSER PRESS, *The Hague, The Netherlands and the Authors*

All the carbon units (AAUs, ERUs, CERs including tCER and lCER, RMUs, EU emission allowances, and verified Emission Reductions, a commodity used for the purpose of selling on to the voluntary market) are all equal to one metric tonne of carbon dioxide equivalent and will all have their own unique serial number. They are first and foremost accounting units, which are tracked and recorded through national registries to be established and maintained by Annex I Parties.

They are tradable instruments which represent an entitlement to release a certain quantity of GHG emissions into the atmosphere; and they are transferable under certain established conditions. Transferability is the most important feature of tradable carbon units as compared to the traditional command and control approach of environmental policy.

However, despite the general similarity of the above-mentioned units, they also differ in significant aspects. They are either created or allocated through a regulating body (international/domestic), and are subject to different legal regimes (mandatory/voluntary); and they differ in the scope of their use, ranging from compliance with international or national legally binding reduction obligations to the fulfillment of voluntary commitments. As a result, they are also embedded in different legal traditions: Civil law and common law systems traditionally define property rights differently – however, applying more or less the same basic criteria-based elements *'usus, abusus, fructus'*[1] to characterize property.

Undoubtedly, each unit creates a right to transfer an entitlement to release a ton of GHG into the atmosphere. Such a right can hardly be classified among the well known exiting rights and may be regarded in most legal systems as a *'sui generis'* instrument.

This paper concentrates on the legal nature of Kyoto units under international and domestic law but it also addresses the legal nature of accounting units created by these policies and measures designed for the implementation of the Kyoto Protocol and which are, as from 2008, backed up by AAUs when transferred from one national registry to another.

Under International Law, the Kyoto Protocol is a 'Treaty' within the meaning of Article 2 paragraph 1, a) of the 1969 Vienna Convention on the Law of Treaties, as it is an international agreement concluded between States in written form and governed by international law, whether embodied in a single instrument or in two or more related instruments and whatever its particular designation.

Therefore, Articles 31 and 32 of the Vienna Convention are relevant to define the legal nature of Kyoto Units. These provisions reflect a combination of methods developed under Customary International Law on the interpretation of Treaties. For instance, several methods of interpretation may be simultaneously used (intention of the Parties, literal/textual approach, teleological approach, effectiveness approach).

Article 31 of the Vienna Convention sets a general rule of interpretation. However, one should note that Treaty interpretation is more concerned with the objec-

---

[1] This maxim refers to the right of an owner to use (*usus*), discharge or transfer (*abusus*), and benefit from (*fructus*) a thing or right he/she possesses.

tive meaning of a Treaty than its application to any particular case. In any case, Article 31, paragraph 1 requires that Parties interpret a Treaty in 'good faith', in the sense that a Treaty must be interpreted in a spirit of loyalty vis-à-vis the rules of the law that are set therein.[2] The interpretation of the terms of a Treaty in accordance with its ordinary meaning is then to be checked against the objective that is pursued by the Treaty. Priority must be given to the terms of the Treaty themselves, including its preamble and its annexes as the foremost reflection of the commitments and will of the Parties. Article 31, paragraph 3 adds that account should be taken of any subsequent agreement regarding the interpretation of the Treaty, while Article 32, paragraph 4 of the Vienna Convention which provides for supplementary means of interpretation that can be used refers to the expression 'special meaning' to cover the case where it is intended that Contracting Parties should provide a special meaning to the expression 'emission reduction units', so that this meaning should apply.

In that respect, the decisions on modalities for the implementation of the Kyoto flexible mechanisms that were adopted as part of the Marrakech Accords recognize that

'the Kyoto Protocol has not created or bestowed any right, title or entitlement to emissions of any kind on Parties included in Annex I.'[3]

This statement means that the Kyoto Protocol does not create any rights to emissions or the atmosphere, but it only creates the right for some Parties to a limited pollution for a defined timeframe. Clearly, the ability of Annex I Parties to trade AAUs does not create public 'ownership' over the atmosphere. In addition, it clarifies that the allocation of AAUs through the quantified targets does not bestow Annex I Parties with any entitlements beyond the first commitment period (2008-2012) of the Kyoto Protocol.

As such, it answers some concerns raised by a number of observers and non-Annex I Parties as to whether the initial allocation to industrialized countries pursuant to Article 3, paragraph 1 would 'grandfather' any future entitlements to emit GHG emissions beyond 2012. In addition, it addresses the question of '*equity*', as the Kyoto regime does not lead to an inequitable 'ownership' of the atmosphere leading to the privatization of the atmosphere for the sole benefit of industrialized countries. In other words, the Kyoto Protocol does not cut the atmosphere into little pieces and distribute it to polluting countries only. In sum, the creation of a trading system under the Kyoto Protocol does not create an 'exclusive right' that affects non-Annex I Parties' right to access the resource.

The atmosphere is a common property of humankind over which no exclusive sovereignty can be ascertained. Therefore, no 'ownership' over the atmosphere ex-

---

[2] Confirmed by the International Court of Justice in Case Concerning the Territorial Dispute Between The Libyan Arab Jamahiriya and Chad (13/02/1994), para. 41 of the Judgement.

[3] Decision 2/CMP.1: Principles, nature and scope of the mechanisms pursuant to Articles 6, 12 and 17 of the Kyoto Protocol.

ists and no State can be barred from access to it. Public international law objects to exclusive sovereignty over natural resources that are common property.

The atmosphere is a natural resource which is open to legitimate but reasonable use, the access to which is open to all.[4] One should keep in mind the UNFCCC Preamble which states that:

> 'Acknowledging that change in the Earth's climate and its adverse effects are a common concern of humankind, (…) States have (…) the responsibility to ensure that activities within their jurisdiction or control do not cause damage to the environment of other States or of areas beyond the limits of national jurisdiction'.

As a matter of International Law, Annex I Parties have committed themselves to reducing their GHG emissions, not to increase them. The Kyoto commitments provide for obligations which are incumbent on States: Governments are allocated AAU as an entitlement to release a certain amount of GHG emissions. States have sovereign rights to use Kyoto units for compliance with the Kyoto Protocol. Therefore, one may argue that States have a public property right to abuse AAUs by selling them under Article 17 of the Kyoto Protocol.

As States have a public property right over Kyoto Units and can trade them, one question is to know whether the legal nature of Kyoto units under International Law could raise some concerns of compatibility with International Trade Law.

In line with the Doha Agenda, WTO and multilateral environmental agreements (MEAs) should be mutually supportive. Article 3, paragraph 5 of the UNFCCC underlines that States' measures to combat climate change should not violate international agreements on trade, in the sense that they must not constitute a means of arbitrary or unjustifiable discrimination or a disguised restriction on international trade. Moreover, Article 2, paragraph 3 of the Kyoto Protocol provides that Annex I Parties shall strive to implement policies and measures in such a way as to minimize, *inter alia*, adverse effects on international trade.

One would argue that trading in Kyoto units does not conflict with WTO law as it constitutes a sovereign exchange of commitments.[5] It can hardly be demonstrated that Kyoto trade-related measures are to be considered as trade measures in the WTO sense. However, domestic measures implementing the Kyoto Protocol could raise WTO concerns and, if falling under the TBT Agreement, may have to be notified (for example, climate-related labeling and product standards).

Are emission allowances 'goods' under the General Agreement on Tariffs and Trade (GATT, 1994)? This question may arise in the context of domestic/regional emission trading schemes, such as the EU ETS, whereby private entities are allocated with tradable emission allowances. The EU ETS directive can be linked to other national or regional emission trading schemes provided that emission allow-

---

[4] Treaties and case law of the ICJ on High Seas Fishing suggest that common spaces must be used in a reasonable/equitable manner so that everyone's right to utilize it is preserved.

[5] Trading in AAUs is about the transfer of emission reduction commitments among Annex I Parties though within the overall cap on Annex I emissions.

ances are mutually recognized. One can argue that emission allowances do not qualify as 'goods' or 'products' under the GATT, as 'goods' are material items with an intrinsic value further to WTO case law.

Are Kyoto units and emission allowances 'services' under the General Agreement on Trade in Services (GATS)? On the one hand, the GATS does not define what a 'service' is. Rather, 'services' are defined in contrast to 'goods/products'. On the other hand, one could hardly argue that Kyoto units and emission allowances are 'services' in as far as they are not activities as such with an economic value, they are listed neither in the WTO's Services Sectoral Classification List, nor in the UN's Provisional Central Product Classification system. However, exchange services (by brokers or through trustee services) on secondary markets on derivatives may fall under the GATS' Annex on Financial Services (financial assets) and would therefore have to comply with basic trade principles of non-discrimination.

Exploring the legal nature of Kyoto units is not just a matter of States' sovereign rights under International Law. In effect, the Kyoto Protocol foresees the participation of private entities, either as authorized entities for trading Kyoto units or as private investors involved in activities leading to the creation of some tradable carbon units, namely JI and CDM credits.

Contrary to AAUs which are allocated to Governments, JI and CDM credits are generated privately, through projects financed by private entities. JI/CDM credits represent a reduction of GHG emissions resulting from a well defined project activity, calculated on the basis of the comparison between the level of verified actual emissions and a counterfactual scenario (defined as baseline scenario). Of private 'origin' these credits are generated by entities as a direct consequence of their investment project activities.

CDM credits are issued *ex nihilo* (reductions achieved in countries where GHG emissions are not capped) by the CDM Executive Board, they are directly delivered to project participants in the CDM registry, and they can be subsequently transferred if the relevant conditions are met (International Transaction Log in place and compliance with Parties' participation requirements).

JI credits are very similar to CDM credits to the extent that they are usually generated thanks to private investments. However, they differ from CDM credits as they are hybrid due to their generation in capped Annex I countries: JI credits are not allocated, but converted from AAUs (which must be cancelled in the host country's registry after their transfer to another national registry so as to avoid double counting among Annex I Parties and compliance with the overall Annex I target of a 5.2% reduction as compared to 1990 levels).

Although JI and CDM credits are primarily to be used by Annex I Parties for compliance with their KP commitments, they may however be used by private entities for compliance purposes at domestic level, as this is the case for the use of JI/CDM credits by operators subject to the EU ETS since the adoption of the Linking Directive.

In sum, one may argue that legal entities have a private property right to abuse JI and CDM credits which result from private investments by selling them to Annex I Parties.

However, the Kyoto Protocol itself does not transfer any direct rights or obligations to entities other than its Parties. Since the Kyoto Protocol does not bestow any rights and obligations on private sector entities, such rights need to be created either through implementation legislation or through single government acts, such as an individual authorization. In order to allow private entities to hold, own and trade units defined under the Kyoto Protocol, including for compliance purposes, these need to be authorized to do so. In that respect, one should keep in mind that private entities must be authorized to participate in JI or CDM projects.[6] Parties can authorize legal entities to use Kyoto units indirectly for complying with their obligation to reduce their emissions through domestic measures implementing the Kyoto Protocol (EU ETS as amended by the Linking Directive[7]).

Independent of whether created through a general act (law) or through governmental decision, with the authorization to participate in a project and to hold and transfer JI or CDM credits, a Government transfers its right to JI or CDM credits to the private entity concerned.

Actually, beyond the sovereign rights of States under International Law, the question of the 'legal nature' mainly arises when legal entities are involved, and depends on the applicable legal regime. The question of legal nature lies with the ownership rights which must be clarified for their treatment under accounting and taxation schemes. *In abstracto*, an allowance represents an asset or (a quasi) property right for the owner if it is transferable. The limits of the property may be defined by the State which may decide that tradable carbon units are state-owned in his jurisdiction, bearing in mind that legal restrictions usually preclude the potential for gains from trade on environmental markets.

In other words, the legal nature and ownership by private entities of carbon tradable units including Kyoto units is to be determined in the light of domestic Law and, more prominently, by contract Law for JI/CDM credits.

If States exercise a public property right on AAUs, one should remember that as of 2008, EU emission allowances are AAUs with an EU tag (a mirror of AAU/a target under the KP). Article 3(a) of the EU ETS Directive defines an 'allowance' as

'(...) an allowance to emit one ton of carbon dioxide equivalent during a specified period, which shall be valid only for the purposes of meeting the requirements of this Directive and shall be transferable in accordance with the provisions of this Directive.'

An allowance is therefore only defined as an entitlement to emit GHG under a predetermined cap on emissions.

---

[6] As for the CDM, the Executive Board has clarified that the written approval to be given in accordance with the CDM rules and modalities constitutes the authorization by a designated national authority (DNA) of specific entities' participation as project proponents in the specific CDM project activity).

[7] Installation operators subject to the EU ETS cannot be allocated EU emission allowances if they are not granted the authorization to emit GHG emissions in accordance with Art. 4 of Directive No. 2003/87/EC which provides for the obligation to surrender a quantity of allowances corresponding to the actual verified GHG emissions.

For legal security and market liquidity, a number of member states have decided to define the legal nature of EU emission allowances through their national transposition measures.

The case of France is interesting in that regard: EU emission allowances were regarded as *sui generis* instruments, though they look very much like a transferable and tradable administrative authorization that can be presented to a successor. The fact that the allowance is only valid for a given period of time does not contradict the perpetuity characteristic usually attached to property rights as the Case Law of the *Cour de Cassation* has ruled that property rights can be exercised over goods for a determined period of time.

The right to transfer the allowance (e.g., the entitlement to emit) having a value (all the more so since allowances are grandfathered), creates a subjective right for the benefit of the holder of the allowance over an immaterial good. Property is materialized when allowances are delivered in the registry. The French measures transposing the EU ETS Directive (Articles L.229-5 to L.229-24 of the Environmental Code) consider that EU emission allowances are goods, while clarifying that they are not financial instruments whereas contracts on futures (derivatives) are. The measures transposing the Linking Directive have also classified the JI/CDM credits as immaterial goods the property over which is recognized when credits are transferred to the registry account of the owner.

An interesting parallel can be drawn with allowances issued under the SO_2/NOx Trading in US Programme. Section 403(f) of the 1990 amendments to the Clean Air Act carefully defines the legal nature of emission allowances as a limited authorization to emit sulphur dioxide, underlining that such allowances do not constitute a property right. However, all normal rights related to property rights (*usus, fructus and abusus*) are available. The US legislator did not want to create property in order to avoid being held liable for compensation in case allowances would be devalued or withdrawn from the market. This has led some authors to describe paragraph 403(f) as 'premised on the confusion between property rights in something and the thing itself'.[8] Indeed, it does expressly define an allowance as not creating a property right, yet allows free transfer and thus recognizes the property rights in the emission allowance, because utilities can receive, hold (i.e., possess), and transfer (i.e., alienate) allowances.

---

[8] See Daniel H. Cole, 'From Local to Global Property: Privatizing The Global Environment? Clearing The Air: Four Propositions About Property Rights And Environmental Protection', *Duke Env L & Pol'y F*, (1999), 10, 103.

MARKETING CERS: LEGAL AND CONTRACTUAL ISSUES FOR SELLERS

Charlotte Streck*

1. INTRODUCTION

As the carbon market continues to develop, there is a need to assist project developers in drafting and negotiating contracts under which they sell the carbon credits they produce. Many - in particular smaller - Clean Development Mechanism ('CDM') projects are being developed by entities which do not have the means to obtain advice from expert law firms. In most cases project developers and sellers of Certified Emission Reductions ('CERs') select carbon buyers solely based on the unit price offered. A few weeks after having communicated their choice to the buyer, they are likely to find an Emission Reduction Purchase Agreement ('ERPA') in their e-mail boxes. There is a fair possibility that the agreement is long, complex and in a language in which the seller is not dominant. Few sellers have legal advisors with previous exposure to CDM projects; many rely on their generalist lawyers or try to manage the negotiation process without legal assistance. Few project developers are represented in forums like the International Emission Trading Association ('IETA') or have the capacity to follow the international discussions in detail.

Project developers that are able to obtain financing for their projects without taking into account a fixed cash flow from the sale of CERs can choose whether to sell their CERs under a forward CER contract (an ERPA) or wait until the CERs are issued and sell them on the 'spot' market. Spot contracts are usually highly standardized and fairly simple documents. For many project developers, however, the forward sale of CERs provides an important opportunity to access a hard currency cash flow and obtain additional upfront financing for the project. Without good advice, this money may come at a hefty price. In many instances contracts under which project developers sell their CERs include delivery guarantees, penalties and strict enforcement clauses. These clauses can easily turn an ERPA from an asset into a liability. They can undermine the usefulness of an ERPA to secure debt fi-

* Dr Charlotte Streck, Director, Climate Focus. The author thanks Robert O'Sullivan, Cathy Lee, Christiana Figueres, Ines Manzano as well as all other participating lawyers and sponsors, in particular the Interamerican Investment Cooperation, for their contribution to the CERSPA initiative (and thus this article). Parts of the description of the CER Sales and Purchase Agreement are taken from the CERSPA Guidance document.

W.Th. Douma et al., eds., The Kyoto Protocol and Beyond
© 2007, T·M·C·ASSER PRESS, *The Hague, The Netherlands and the Authors*

nancing and put an entire project at risk, as well as the company behind it. Many carbon contracts also lack elements which are common to other long-term purchase agreements, such as price and inflation adjustment clauses, or limitations on liabilities.

Taking into account the economic opportunity and the environmental benefits of the numerous small and medium-sized CDM projects that are currently under development as well as the risk exposure of sellers under carbon contracts, it is important that sellers of CERs gain access to legal CDM intelligence. This paper seeks to show how CDM carbon contracts can mitigate risks taking into account the particularities of their project. The paper will give a brief introduction to the main issues to be considered in the negotiation of a CDM forward contract. It will also briefly summarize the main features of the existing model CDM ERPAs, and present a new template for an Emission Reduction Sale and Purchase Agreement ('CER SPA') currently under development. The paper will draw on the practical experience of its author as well as the discussions and exchanges of views in the context of an international initiative to develop the CER SPA which will officially be launched on the occasion of Carbon Expo in Cologne, Germany on 4 May 2007.

2. CDM MARKETING STRATEGY: MAIN ATTENTION POINTS

2.1 Definition of a sales strategy

Carbon contracts that regulate the transfer of emission reductions under the CDM have to be innovative and robust. They must support the contractual parties' efforts to successfully implement their cooperation long after the attention of initial project developers and consultants has moved on to other tasks. Carbon contracts define the relationship between parties in an emerging market which is characterized by a wide variety of uncertainties and risks. The contracts need to record the agreement between the parties, identify responsibilities, allocate risks, establish rights, and create clear and enforceable obligations.

Before soliciting proposals from potential CER buyers, a project developer and carbon seller needs to carefully evaluate the available options to the market and sell the CERs it expects to generate. This is often not an easy thing to do, as many project developers are small to medium-sized entities far away from sophisticated markets and the centres of carbon expertise. While the internet holds a wealth of information on CDM, carbon markets and prices, not all sources of information are reliable. Further, much of what can be found is contradictory or insufficient to build a solid understanding of the market. To date, the carbon market is not a homogenous market, but a mosaic of various markets differing in their scope and coverage, their terms and conditions, their linking to the international regime, and their voluntary versus obligatory nature. Prices offered for carbon credits differ according to the various segments of the market, the risk associated with the delivery of the credits (which includes project, regulatory and counterparty risk), the timing of the delivery, and the general demand and supply of credits. It is therefore important to put any offered carbon price into perspective. While the price a seller can obtain for CER is crucial, it is important not to neglect the other terms and conditions of an ERPA. If a CDM project is financed and carries a low implementation risk, a project developer may be best advised to wait with the marketing of the CERs until the project is registered or even until the CERs are issued. If the project developer looks for an advance payment or technical assistance, the project developer is advised to choose a CER buyer who has some technical understanding of the project – and potential delays and hurdles in its implementation. It is worth spending some time studying the full implications of the timing of the CER sale. Independent consultants (without an interest in a particular sales strategy), banks and legal experts may help highlight risk areas and financial implications.

A first guidance in defining a marketing strategy is provided by the following checklist:

Strategic checklist for CDM project developers

Is the CER cash flow essential for my project?	If the project depends on the CER cash flow the seller is more likely to be risk adverse. The seller will seek to sell at least a part of the future CERs under a forward fixed-price contract. The revenue expected under such contract should be sufficient to cover the investment and operating costs of the project.
Do I need additional funds to cover my investment costs? Do I need an advance payment?	In case the seller is lacking upfront investment capital, it should find a buyer willing to advance some of the funds. In this case the seller should evaluate whether it is able to give a guarantee for the advance payment (parent/performance guarantee or a letter of credit).
How would outsiders value my company? Do I have any credit rating or a balance sheet backing the transaction?	The credit rating and reliability of the seller will influence the buyer's purchase decision, in particular if there is an advance payment involved.
Would I benefit from technical advice and knowledge transfer?	In case the seller is looking for advance and technical assistance, it should consider choosing a buyer which is knowledgeable about the project class and technology involved. When providing advice and reviewing the project design and implementation, the buyer will be more easily prepared to share some of the project risks.
Which are the project-related risks of my project?	A careful evaluation of risks is important as risk is reflected in the project's ability to generate CERs. If the project is considered to be very risky, it is important to be conservative in the

CER projections and not to issue delivery guarantees. Project-related risks include the risk of obtaining financial closure, technical and project design risks, supply and construction risks, regulatory and political risk.

Which are the CDM-related risks of my project?

The development of a CDM project adds another layer of risks to the project that needs to be taken into account. Such risks include the validation and registration risk, the baseline risk, the risk of a highly political regulator (the Executive Board).

How significant is the yield in CERs of my project? How many CERs will I have to sell every year and over the crediting period of the project?

The size of the project and the number of expected CERs is also an important factor when establishing a CER sales strategy. The smaller the project and the fewer the number of CERs, the easier should be the transaction. Bigger projects can adopt more sophisticated sales and marketing strategies combining secure with risky transactions and enter into more than one contract.

Is this my only CDM project or do I have more projects in the pipeline?

If a seller has several small projects he should evaluate the possibility to bundle these projects in one CDM project or under one contract. In case a developer has a portfolio of several medium sized-big projects, he can choose to pursue different sales strategies for each project. The developer can also pool projects and sell CERs from the pool thereby hedging the delivery risk of the individual project.

Do I have full market access?

The degree of market access a seller has often determines the marketing strategy. In case the seller has no full-time personnel working on the CDM or it is too far from the centres of the international carbon market, it may consider getting professional advice in marketing his CERs.

What price do I expect and which price can realistically be achieved?

Consideration of the different points listed will help the seller to position his project in the market. Based on a careful evaluation the seller can develop a term sheet which will be circulated among potential buyers. The price for which a seller will be able to sell CERs will depend on the risk profile of the project as well as on the negotiation skill of the carbon seller.

2.2   Timing of the sale: spot and forward markets

To date, most project developers sell CERs under forward contracts. Forward contracts are contracts under which a seller agrees to deliver a specific commodity to a
buyer at some point in the future. Unlike futures contracts (which occur through a
clearing firm), forward contracts are individually negotiated and not standardized.
The first forward carbon contracts were developed by pioneering buyers that started
to show an interest in purchasing forward CERs as early as 1999. These first buyers
were government purchase programmes and carbon funds managed by the World
Bank. This class of buyers sought (and still seeks) to manage a long-term carbon
liability rather than to satisfy a short-term (annual) compliance need. Back then, the
carbon market was still in its infancy and traders or retailers of credits were almost
absent. Buying carbon credits was an esoteric and risky proposition and few private
sector entities believed in the future of the market. The primary objective of the
early institutional buyers was to help develop a market that did not previously exist
in the face of significant risks and purchase cost-efficient emission reductions to be
delivered sometime before 2013. Sellers, on the other hand, had little confidence in
the market and the future value of carbon credits. They were interested in signing
contracts early and selling the future stream of carbon credits for a fixed price.
Their main concern was to recover their project development cost hoping that future emission reductions would be assigned some value. Interests of buyers and
sellers therefore converged when they entered into long-term off-take agreements
applying fixed prices for CERs or verified emission reductions ('VER') delivered.[1]
These first agreements became known as Emission Reduction Purchase Agreements
(ERPAs) and set a standard for carbon transactions.

The entry into force of the Kyoto Protocol and the implementation of various
emission trading schemes, most importantly the EU Emission Trading Scheme, led
to increased private sector interest in buying CERs. Today, the CDM market is
dominated by private players. This 'privatization' of the market went hand in hand
with an increasing sophistication of contractual structures. A market in carbon credit
notes, options and other securities emerged and complemented the more traditional
forward purchases commonly used by institutional buyers. CDM project developers can now choose the best conditions and prices for their projects, comparing a
wider array of different offers.

In this environment sellers have started to appreciate risk and opportunity. When
selling under forward contracts, pure price promises are no longer enough. Instead
sellers gained confidence and started negotiating upfront payments, flexible price
structures and price reopeners, technical assistance and payment guarantees.[2] The
bigger number and greater variety of buyers (large utilities, banks, investment funds,
traders and speculative funds next to compliance funds, government buyers and

---

[1] Under VER contracts the buyer assumes the additional risk to turn the emission reduction into
a CER, thereby assuming the international approval, registration and issuance risk.

[2] Carbon buyers like the World Bank had offered some of these services and contractual features previously (e.g., technical assistance and upfront payments).

international organization-managed carbon funds) made it possible to find the appropriate offer and entity matching the expectations and needs of each type of seller.

However, most importantly project developers and sellers started to hold back credits to await further increases in price. Today, more and more sellers decide not to sell CERs under a forward contract but rather to sell CERs on the 'spot' market. This is a risky proposition, in particular since international carbon markets are characterized by high price volatility. In May 2006, the price of EU allowances[3] fell by half within a week.[4] Current forecasts about future price developments are mixed, with a tendency towards predicting stabilization. Project developers that hold back their CERs will benefit from higher prices for a more secure product (CERs without project risks attached), provided that the price for CERs does not drop below the price offered under a forward contract. They thus run the risk of not being able to cash in on this advantage if prices drop further. If prices rise, however, their benefit will be greater. Trading on the spot market and price speculation may therefore pay off for those that can afford to speculate.

CERs traded on the spot market are ideally free from all project and CDM risks. Under spot contracts purchase and settlement coincide within a couple of days. If settlement takes longer than a week, the contracts move again towards forward contracts. Since CERs once issued in an Annex I[5] registry are a (relatively) homogenous commodity, contracts that deal with the purchase and sale of these CERs can be simple and standardized.

It is important to note, however, that despite the increasing confidence among project developers, an imbalance between entities from industrialized Parties to the Kyoto Protocol (referred to as Annex I countries) and entities from developing Parties (referred to as Non-Annex I countries) persists. Non-Annex I entities do not have the same access to the market as Annex I entities. Most importantly, private entities from developing countries do not have access to trading accounts in emission registries in their own countries. The CDM registry in which Non-Annex I entities may hold accounts is limited in its features. CDM registry accounts are merely holding accounts that do not allow the receipt of credits from anywhere but the CDM registry's pending account. Additionally, private sub-accounts in the CDM registry are created under international law which does not extend legal protection (against, e.g., national claims, confiscation, expropriation) to private entities to which certain accounts have been assigned.

At the moment, few developing country entities have access to Annex I registry accounts. They therefore sell or plan to sell the future yield in CERs in forward contracts or directly from the CDM registry's pending account. Since project developers neither control the pending account nor the host country accounts in the CDM registry, the commitment of Non-Annex I carbon sellers to make timely trans-

---

[3] EU allowances are the emission rights allocated by European governments under the EU ETS.

[4] CCX Carbon Market, May 2006 Trading Summary. <http://www.chicagoclimatex.com/news/newsletters/CCX_carbonmkt_v03_i05.pdf>.

[5] Annex I of the UNFCCC and the Kyoto Protocol list those countries with a quantitative emission reduction obligation under the Kyoto Protocol.

fers of CERs is risky. The current CDM infrastructure hampers the possibility for Non-Annex I entities to determine the best moment for selling carbon credits. Any transaction has to be carried out by the CDM registry administrator instead of being under the direct control of the Non-Annex I entity. Sellers can therefore hardly commit to deliver CERs on particular dates. Unless sellers and project developers open and manage accounts in an Annex I (and fully operational trading registry) or enter into a trust arrangement with an Annex I entity, their ability to participate in spot arrangements remains imperfect.

In case the project is burdened with uncertainties, the project developer should consider selling VERs instead of CERs, effectively shifting the regulatory risk to the buyer. There are a few institutional buyers, such as the World Bank, that offer to buy VERs and convert them into CERs. Outside of the regulated markets, the project developer can also sell credits to the so-called voluntary or retail market.

In case the project developer chooses to sell CERs in order to benefit from the higher price offered for those credits, it should be careful in accepting delivery guarantees or penalties, in particular if it is new to the market and without concrete CDM project experience. In this case, it can also negotiate for technical assistance that helps it to maximize the CER yields from its project. A number of CER buyers combine the off-take of CERs with the offer to provide technical advice to the project. Such offers are not only interesting from the point of view of mitigating delivery risks, they are also interesting as they help to cement new partnerships and cooperation, which significantly reduces the risk of disputes between the supplier and the off-taker of carbon credits.

3. EXISTING CDM CARBON CONTRACT MODELS

When the pioneers of the CDM, the World Bank through its Prototype Carbon Fund and the Dutch Government through its Certified Emission Reduction Unit Purchase Tender ('CERUPT'),[6] started operationalizing the CDM, they could not revert to existing contractual models. Article 12 of the Kyoto Protocol was the sole source of knowledge and as such was silent with respect to the way Annex I and Non-Annex I entities should cooperate in the implementation of the mechanism. Many negotiators had expected that the CDM would trigger direct investments by Annex I entities into projects. However, as it turned out the assignment of value to carbon credits was easiest to be achieved under a sales-and-purchase transaction. Consequently, the first template carbon contracts where developed as emission reduction purchase agreements (ERPAs). Since then ERPAs have become the standard format for forward purchases of CERs.

The World Bank and IETA have developed template ERPAs that are publicly available.[7] The pioneering work and subsequent revisions by these organizations

---

[6] The Dutch Government's first tender rounds targeted JI projects (ERUPT). CERUPT followed later and was modeled after the successful experience of earlier rounds of ERUPT.

[7] <http://carbonfinance.org/Router.cfm?Page=DocLib&ht=34&dtype=102&dl=0>.

have produced sophisticated legal documents that summarize much of the existing intelligence in this emerging legal discipline. Both models have been developed by institutions that represent mostly buyers ('IETA') or act as buyers themselves (the World Bank). The authors of both precedent agreements have made an effort to develop balanced and fair documents serving both the purposes of the buyer as well as the buyer. However, both documents are in structure and language not easily comprehensible to project developers without experience in international carbon transactions.

3.1 The World Bank ERPA

The actual World Bank ERPA is based on 'General Conditions'[8] and a CDM CER Purchase Agreement.[9] The development of General Conditions for carbon transactions reflects the general practice of the Bank to rely on a set of non-negotiable conditions for its transactions. General conditions exist for almost all financial instruments and contracts offered by the World Bank. The advantage of such a system is that it guarantees transparency and consistency among contracts signed by the World Bank. It reduces costs and facilitates management approval of the transactions, on one hand, and implementation and compliance monitoring on the other. The reliance on General Conditions is in particular advantageous where a large number of agreements of the same type are entered into. The disadvantage of the approach is that the possibilities for the World Bank's counterparty to negotiate alternative arrangements are limited. But then again, World Bank contracts are often not subject to commercial negotiations in the private sector sense of the word anyway. The World Bank CDM ERPAs[10] are simple documents which summarize the commercial details of the transaction. The General Conditions apply to the ERPA to the extent that the specific agreement does not include any variation of the Conditions.

3.2 The IETA ERPA

The basis of the IETA ERPA is a Code of CDM Terms Version 1.0 (2006) (the Code).[11] The IETA ERPA is a simple document which incorporates in its Schedule those provisions of the Code that the parties agree on and consider appropriate. The Code is an extensive document which includes numerous drafting suggestions for various articles and clauses. Those drafting an ERPA can help themselves among the clauses offered and thus construct their particular and individualized ERPA. The definitions included the Code are referred to in the ERPA by default. The IETA Code and ERPA follow the system of ISDA Master Agreements with their set of basic agreed and non-negotiable definitions and their system of annexes, which

---

8 <http://carbonfinance.org/docs/CERGeneralConditions.pdf>.
9 <http://carbonfinance.org/Router.cfm?Page=DocLib&CatalogID=28153>.
10 The World Bank offers VER as well as CER ERPAs.
11 <http://www.ieta.org/ieta/www/pages/getfile.php?docID=1794>.

summarize the commercial details of the transaction.[12] The underlying idea of creating a format of core provisions to which the actual details of the transaction can been annexed has the appeal that if such contract models are widely used, they increase the efficiency of the negotiations significantly. Such contractual models are particularly useful where parties intend to enter into multiple contracts. The objective of the Code is to standardize as many of the CDM ERPA provisions as possible.

Initial efforts to use the IETA ERPA have however revealed difficulties in the use of the agreement. The system of the IETA ERPA produces a short, albeit complicated, agreement that even for lawyers is difficult to deal with. While the provisions of the Code are likely to find their way into project-specific ERPAs, it is rather unlikely that the IETA ERPA as such will find many users.

The World Bank ERPA is simpler and more accessible than the IETA agreement. Instead of leaving it up to the user to build his agreement from a warehouse of provisions, the World Bank agreement incorporates the full text of the General Conditions into the ERPA. The agreement, however, is nevertheless unlikely to determine future business practice. The General Conditions are World Bank-specific and few private sector entities will agree to incorporate the World Bank's General Conditions without further modifications into their agreements. Similar to the IETA contract, they may however include particular provisions into their agreements.

The World Bank and IETA's efforts to standardize and publish CDM ERPAs have provided the carbon market with important model documents. The World Bank's main goal in preparing a template agreement was to standardize its own carbon transactions emission. IETA had a more ambitious objective when it tried to develop a standardized agreement for any CDM transaction. Experience shows, however, that CDM transactions are somewhat resistant towards standardization. CDM projects come in all kinds of forms, sizes, and varieties. Regional context, project type, project size and the financial status of the project and the project owner determine the conditions for and the format of forward CER transactions. Following the needs of the project, agreements governing the transfer of CERs on the primary market continue to show a great variety. Unlike spot and other transactions on the secondary market, it is unlikely that the primary CER market will soon rely on a standardized ERPA, or that a standardized agreement would necessarily reflect the needs of the current CDM market.

4.   CERSPA: A NEW OPEN-SOURCE CDM CONTRACT

While the standardization of ERPAs is unlikely to happen soon, there is an unquestionable need for specialized legal advice in the CDM market. In particular carbon sellers are often still excluded from proper legal intelligence. Most project develop-

---

[12] International Swaps and Derivatives Association.

ers still find themselves alone and exposed in ERPA negotiations with carbon buyers that insist on using their contractual models. To address this situation, more than twenty lawyers, the overwhelming majority of whom are from developing countries, have come together in a carbon contract drafting initiative that has developed a resource document and contract template which is aimed at assisting project developers in CDM contract negotiations. The objective of the initiative is therefore to develop a simple, understandable, and balanced CDM project agreement. The agreement is going to be supported by publicly available open-source legal documentation that will assist sellers in participating in the international carbon market on an equal footing with more experienced buyers. Input from the financial sector should enable the resulting template to be used more effectively to attract debt financing. The model agreement together with the supporting documentation will be made readily available to the public in English, Spanish, Portuguese, Chinese and French. The agreement and supporting documentation are expected to continue to evolve over time as more lawyers, project developers and other stakeholders have the opportunity to use it and improve the agreement by adding their input and expertise.

The CERSPA does not prescribe particular contractual clauses. Instead it aims to identify key issues for sellers and financial institutions and provides them with a tool-box of legal ideas and drafted language that will help CER sellers to realize the value of their CERs without putting their project at risk. While it is hoped that such a document may form the basis of subsequent agreements with medium and smaller-sized buyers, it is unlikely that larger institutions will abandon their preferred precedents. An alternative model carbon contract will nonetheless prove to be a valuable resource to sellers in these instances, as it can be used by sellers to understand key issues and to identify the alternatives that are available.

4.1 Contractual cornerstones

The following section shows some of the clauses and explanations as produced by the CERSPA initiative. The full documentation will be available from May 2007.[13]

Sale and purchase of CERs

CERs are the result of the emission reductions generated by a registered CDM project and defined in the decision of the COP/MOP[14] of the Kyoto Protocol. Legal experts are still discussing whether CERs legally constitute a commodity, a security, a permit or another type of intangible right. Few national laws contain clear definitions of CERs or other carbon rights. In some jurisdictions the legal nature of CERs depends on the context and the same CER can be characterized differently for tax, accounting, financial regulation, and contractual purposes. There is how-

---

[13] A link will be included in the Climate Focus website <www.climatefocus.com>.
[14] Conference of the Parties to the UNFCCC serving as Meeting of the Parties to the Kyoto Protocol.

ever no doubt about the fact that CERs can be traded. The establishment of the price and volume of CERs sold stands at the centre of most CDM projects and all ERPAs. There are several ways to establish the price for CERs in an ERPA:

Fixed price. The simplest approach is a fixed unit price per CER. This fixed price will remain in effect for the term of the agreement. This approach provides certainty to both seller and buyer. It does not take inflation or market fluctuations into account, however.

Indexed Price. An indexed price will refer to a spot market price to calculate the unit price per CER. As a result the unit price will fluctuate and change with each annual payment. Such a calculation method entails opportunity and risk for both the seller and buyer, depending on how the reference spot market price changes over the term of the agreement. Using a simple indexed price means neither the seller nor its banks will be able to calculate the carbon revenue, and thus the value of the agreement.

Combination of indexed and fixed price. Combining a mixture of a fixed price and an indexed price referring to part of the unit price guarantees a minim price and reduces the impact that spot market price fluctuations has on the unit price. Such mixed formula can also be combined with a ceiling or cap on the unit price similar to that described below.

Indexed price with a floor and ceiling. Including a floor and a ceiling or cap on the unit price protects both the seller and the buyer from larger movements in the spot market price and should assist in longer-term planning.

There are different methods for calculating the spot price. One can choose to use the spot price for CERs over the last 12 months prior to delivery which will reduce the risk of being exposed to temporary or short-term market fluctuations prior to delivery, but as a result may not accurately reflect the latest market position. Alternatively, one can use the spot price two business days prior to delivery as the reference price and thus ensure that up to date prices are used.

Regarding the volume of CERs, there are two commonly used methods of describing how many CERs are to be bought and sold: fixed amount and percentage volumes. Fixed amount contracts are more common as they clearly establish the minimum/maximum purchase liability for the buyer and the amount of revenue that can be expected by the seller. Contracts establishing percentages of the CERs generated as sales volumes have clear advantages for the seller, since the seller can meet percentages of delivered (up to 100%) without accepting annual minimum delivery amounts.

Where a contract only covers CERs generated during the first commitment period of the Kyoto Protocol (2008-2012), the period in which CERs are to be generated ends on 31 December 2012. If CERs generated during a possible second commitment period are also being bought and sold a later date applies.

Delivery

Care must be taken in determining when delivery is to take place. Most contracts contain annual delivery obligations but shorter or longer time periods are also possible under the CDM rules. Compliance with the EU Emission Trading Scheme commitments are assessed on 30 April each year, so most EU buyers request delivery early in the year. However, if the date is set too early (for example January) it may not leave enough time have the CERs generated in December verified, issued and then delivered to the buyer.

Determining where delivery is to occur is principally influenced by how 'Delivery' is defined. However, whether or not the buyer is listed as a project participant and who has communication rights with the Executive Board can also affect where and how CERs are delivered. In the proposed CERSPA, delivery is defined as the receipt of CERs into a registry account nominated by the buyer. If the buyer does not have a registry account that can accept the CERs or fails to nominate an account, delivery has occurred with the issuance of the CERs into the pending account of the CER registry. In most cases the buyer will be listed as a project participant. This allows CERs to be forwarded directly into the buyer's registry account – either a temporary holding account in the CDM registry if the international transaction log is not operational, or a national registry account if the international transaction log is operational. If the buyer is not a project participant then the CERs will be first forwarded into an account of the Seller or a third party before they are transferred to the buyer. The international transaction log must be operational for this to occur. The CERSPA states that the seller has sole communication rights with the EB. If this is amended to grant joint communication rights to the buyer or seller both parties may need to sign instructions telling the Executive Board where to forward the CERs. If the buyer has communication rights the seller should also avoid the possibility that it can only meet its delivery obligations if the Buyer first undertakes certain actions, such as telling the Executive Board to forward CERs into its registry account.

Communication with the Executive Board

Entities listed as project participants in the project design document have the right to communicate with the Executive Board on a number of project-related issues, including where to distribute the CERs once issued if they are listed as the focal point for communication. Project participants are able to receive CERs directly when they are issued.

There are a number of different ways the parties can communicate with the Executive Board; one party can have sole communication rights for all aspects of the project; the parties can have joint communication for all aspects of the project; one party can have sole communication rights over some aspects of the project, but the parties have joint communication over others, such as distribution of CERs.

Whether or not the buyer has provided any financing towards the project and whether the seller is selling 100% of the CERs to the buyer should be taken into consideration when negotiating the communication rights.

The seller should aim at retaining the sole communication rights with the Executive Board in order to avoid giving the buyer the power to determine (and delay) the process of delivering CERs. If joint communication is chosen the buyer will need to sign any communication sent to the Executive Board for the delivery of CERs to occur. Similarly, if the seller intends to sell some of the CERs to third parties, the seller's ability to deliver CERs to the third party will also depend on the buyer's joint signature of the communications with the Executive Board. The parties should agree to amend the communication rights to remove any rights of the buyer once all their CERs have been delivered.

4.2   Events of default and remedies

The defaults and remedies are a key aspect of the agreement that affect whether or not the agreement becomes an asset to the seller that can be used to attract debt financing or a liability that has the potential to bankrupt the seller. The CERSPA aims at minimizing liability for both parties unless there has been an intentional breach or gross negligence.

The agreement lists a number of standard events of default which are applicable to both parties. These events include:

(a)   material breach of the agreement;
(b)   misrepresentation;
(c)   insolvency and bankruptcy.

'Payment Failure' is an additional event of default by the buyer while 'Delivery Failure' is defined as an additional event of default by the seller. Delivery Failure refers to delivering both an annual guaranteed amount and a total amount by set dates. If the seller has a delivery obligation to deliver CERs above the guaranteed amount each year (e.g., to deliver everything that is produced), it is possible to amend the definition of Delivery Failure to include cumulative guaranteed amounts for each year. This can be structured so that an over-delivery in previous years will compensate an under-delivery in later years. This will reduce the risk that a low production year will result in a delivery failure. This formulation can replace the annual guaranteed amount concept or act as an additional trigger.

The buyer's and seller's remedies are split into two types: remedies for regular events of default, and stricter remedies for defaults caused by intentional breach or gross negligence. The main remedy in the case of a regular event of default is the right to terminate the agreement. In case of willful misconduct or gross negligence, the affected party can claim damages.

Stricter remedies for Delivery Failure and Payment Failure are often considered even if the shortfall in delivery is not due to intentional breach or gross negligence.

These tend to include obligations to deliver replacement CERs or pay damages if there is a Delivery Failure. The risks and potential benefit of these remedies to each party need to be considered as they increase the risks for the seller which should be reflected in a higher unit price. Lower delivery default amounts (minimum annual CERs to be delivered) and cumulative delivery defaults can also be considered. In return it is not unreasonable for the seller to expect stricter remedies for Payment Failure, which could include the payment of damages pegged against the market price. The buyer needs to assess the potential benefits gained by having stricter remedies for Delivery Failure in light of the time and costs of enforcement.

5. CONCLUSION

The CDM offers project developers access to new financing sources in hard currency. While in the early days of the CDM the market was dominated by Government and institutional CER buyers, the last two years have seen an increased interest among private entities in participating in the market. Funds and purchase programmes have mushroomed throughout Europe, Japan, and – to a lesser extent – in North America. The increasing number of private entities entering the market has led to a higher variety of transactional structures to be used. In view of this increasing complexity, the World Bank and IETA have tried to compile the existing legal intelligence in contractual models and General Conditions/Code of Terms. The objective of both institutions was to lower transactional costs and increase the transparency of the market. Building on the experience of an international group of legal experts working in the field of climate change, the CERSPA initiative is aimed at developing a contractual model (and an explanatory guidance document) taking account of the needs of small and medium-sized carbon sellers. The CERSPA is designed to assist project developers in negotiating robust, fair and equitable carbon contracts.

Part 2

EXPERIENCES AND PERSPECTIVES IN THE
INTERNATIONAL CLIMATE REGIME

THE COMPLIANCE REGIME OF THE KYOTO PROTOCOL

Massimiliano Montini*

1. INTRODUCTION: DISPUTE SETTLEMENT AND DISPUTE AVOIDANCE UNDER MEAS

In the last few decades the astonishing development of international environmental law, which has occurred mainly through the conclusion of an huge amount of treaties among sovereign States regulated under international law, commonly referred to as multilateral environmental treaties (MEAs), has been coupled neither with a record of a high level of compliance with such agreements nor with the parallel development of effective systems to promote and facilitate adequate compliance with their provisions.[1]

I have been exploring elsewhere the essential characteristics of this phenomenon, while trying to assess in particular why in the case of many international treaties in the field of environmental law traditional dispute settlement regimes normally employed under international law do not seem to be suitable to deal with emerging disputes. Several reasons have been detected for such a phenomenon, which can essentially be grouped under two headings: a lack of will and a lack of resources.[2]

In this respect, it ought to be recalled here that the evidence collected during the operation of many MEAs has shown that in most cases the reasons for non-compliance by a Party to an environmental treaty are related to the lack of adequate technological or financial assets, rather than to the explicit desire not to comply with some specific international obligations stemming from an international agreement voluntarily agreed upon by a State Party. Therefore, the issue of how to promote compliance with MEAs more efficiently essentially lies in the efforts which can be deployed in trying to develop specific 'compliance regimes' for such international treaties, which are essentially tailored to the specific problems that a Party may presumably encounter in the implementation and application of the provisions of

---

* Associate Professor of European Union Law, University of Siena, Italy.

[1] See M. Montini, 'Improving Compliance with Multilateral Environmental Agreements through Positive Measures: the Case of the Kyoto Protocol on Climate Change', in A. Kiss et al., *Economic Globalization and Compliance with International Environmental Agreements*, Kluwer Law International, The Hague, 2003, p. 158. See also the specific literature cited there on compliance with MEAs.

[2] Ibidem, p. 159. See also M. Bothe, 'The Evaluation of Enforcement Mechanisms in International Environmental Law', in R. Wolfrum (ed.), *Enforcing Environmental Standards: Economic Mechanisms as Viable Means?*, Berlin, 1996, p.13.

W.Th. Douma et al., eds., The Kyoto Protocol and Beyond
© *2007, T·M·C·ASSER PRESS, The Hague, The Netherlands and the Authors*

that specific treaty, or in other words in the fulfillment of the obligations stemming from such a treaty.

A new category of dispute settlement regimes, commonly called 'compliance regimes', has therefore emerged in the last few decades, particularly in connection with the operation of MEAs. The main common feature of such legal systems is represented by their essential nature of dispute avoidance rather than dispute settlement regimes, in the sense that most of the provisions and related procedures developed thereunder refer to the main basic aim of trying to understand the reasons for non-compliance, by collecting the relevant data through a well-organized information system, in order to be able to help a non-complying party, mainly through the direct provision of (or the facilitation of the conditions for receiving) the necessary technological or financial assistance to resume compliance with the violated international norms at stake.

In fact, compliance regimes are normally equipped to promote the resumption of compliance through the adoption of soft consequences, such as the promotion of capacity building, technology transfer and the provision of financial assistance, often grouped under the heading of 'carrots', as compared to the harder consequences, commonly called 'sticks', which may be represented by the suspension of certain advantages granted by a certain treaty or the suspension or termination of the treaty itself against the non-complying Party as a consequence of a material breach, as provided by the Vienna Convention on the Law of the Treaties.[3]

Since compliance regimes developed under most MEAs are normally neither equipped, nor really focused on the possibility to issue real and effective sanctions against non-complying Parties, one of the issues to be primarily addressed when trying to assess the scope and reach of MEAs is the one pertaining to the relationship of the compliance regimes developed under such environmental agreements with the existing traditional dispute settlement provisions operating under public international law.[4]

In this sense, in particular, the main questions to be answered are the following: first of all, are these compliance regimes of an exclusive nature with respect to the disputes which may arise under the provisions of the treaties to which they refer and can they therefore exclude, or not as the case may be, the application of traditional dispute settlement procedures and rules normally available to Parties under international law? In addition, if the answer to the first question is in the sense of the non-exclusive nature of such regimes, is there any 'priority' for the application of the pertinent compliance regimes over the traditional dispute settlement procedures and rules normally available to Parties under international law?

---

[3] Ibidem, p. 160.

[4] See M. Koskenniemi, 'Breach of a Treaty or Non-Compliance? Reflections on the Enforcement of the Montreal Protocol', in *YIEL*, 1992, p. 123; M. Fitzmaurice & C. Redgwell, 'Environmental Non-Compliance Procedures and International Law', *NYIL*, 2000, p. 43 ff.; G. Loibl, 'Environmental law and non-compliance procedures : issues of state responsibility', in M. Fitzmaurice and D. Sarooshi, *Issues of state responsibility before international judicial institutions*, Oxford, 2004, p. 213; U. Beyerlin, P.T. Stoll, & R. Wolfrum (eds.), *Ensuring Compliance With Multilateral Environmental Agreements: a Dialogue Between Practitioners and Academia*, Martinus Nijhoff, 2006.

As far as the first question is concerned, under the applicable rules of international law nothing in principle may prevent the Parties to an international agreement from developing specific rules and procedures to settle their disputes under the said agreement. However, it is to be assumed that in no case may a non-compliance regime, such as one of those developed under the existing MEAs, exclude the application of the traditional dispute settlement provisions normally available under international law. In the case of the Kyoto Protocol (KP), in particular, this is also confirmed in Article 19 which refers to Article 14 of the United Nations Framework Convention on Climate Change (UNFCCC), according to which:

'In the event of a dispute between any two or more Parties concerning the interpretation or application of the Convention, the Parties concerned shall seek a settlement of the dispute through negotiation or any other peaceful means of their own choice.'

As a consequence, no doubt may be said to exist concerning the non-exclusive nature of the compliance regimes developed under MEAs in general and the Kyoto Protocol in particular. However, the problem of the relationship between these two sets of systems under international law still remains with regard to, in particular, the possibility to determine a 'priority' for the application of the compliance regimes specially designed for a certain MEA over the traditional means of dispute settlement available under international law.

The issue is too complex to be extensively analyzed here. However, in very general terms, it may be said that the recognition of a 'priority' for the *ad hoc* compliance regime over the traditional means for the peaceful settlement of disputes under international law would be very difficult to justify, at least in the absence of a specific provision in this sense contained in the treaty at stake. Obviously, the risk inherent in such an approach is that a non-complying Party to a MEA for which a specific compliance regime is in place may be subject to the parallel application, possibly even in competing and conflicting terms, of both systems, namely the compliance regime on the one side and one of the traditional dispute settlement means on the other.

In this respect, a particularly difficult situation may arise if a non-complying Party to a MEA, as a consequence of its wrongful behavior, is subjected to suspension from some of the rights and privileges arising from the treaty at stake, or, in even stricter terms, the treaty itself is suspended against that Party in response to a 'material breach' of its provisions, pursuant to Article 60 of Vienna Convention on the Law of Treaties.[5]

In the case of the compliance regime elaborated for the Kyoto Protocol, in particular, suspension from participation in the flexible mechanisms is one of the consequences which may be applied by the Compliance Committee against a Party which does not comply with the eligibility requirements established for the said

---

[5] See M. Fitzmaurice, 'The Kyoto Protocol Compliance Regime and the Law of Treaties', in M. Fitzmaurice & O. Elias, *Contemporary Issues in the Law of Treaties*, Utrecht, 2005, p. 289 ff.

mechanisms. Such a consequence certainly represents an interesting example of suspending a Party from some rights and privileges arising under a treaty, which may be applied within the framework of a compliance regime. Moreover, it has the positive effect of leaving the treaty in force with respect to the non-complying Party, which is particularly relevant for MEAs, where, as it has been aptly explained, the suspension or termination of the treaty against a non-complying Party may not necessarily serve the interests of the global environment.[6]

This leads us to consider the issue of the possibility to suspend in whole or in part the application of a treaty towards a Party not in compliance with its obligations under that treaty. The prerequisites of such a suspension are determined by Article 60 of the Vienna Convention on the Law of Treaties and should amount to a 'material breach' of the treaty provisions. In our context a material breach, as defined by Article 60, consists of 'the violation of a provision essential to the accomplishment of the object and purpose of a treaty'. What exactly triggers the 'material breach' must be obviously assessed on a case-by-case basis, but for the Kyoto Protocol, reference should probably be made to at least those types of breaches in response to which the Enforcement Branch may intervene on the basis of the compliance regime elaborated under the Kyoto Protocol.[7]

In the present case, there is still some doubt as to whether the Compliance Committee, through the Enforcement Branch, may proceed in some specific circumstances to suspend the Kyoto Protocol as far as a non-complying Party is concerned. However, despite the fact that the compliance regime elaborated under the Kyoto Protocol may be considered as a sort of '*lex specialis*' with respect to the provisions of Article 60 of the Vienna Convention, it seems to me that nothing in principle may exclude the possibility to suspend the treaty in respect of a non-complying Party if that Party is guilty of a 'material breach', even if such a possibility is not explicitly foreseen in the Kyoto Protocol compliance regime. It remains to be seen, however, what the relationship of such a suspension could be with the consequences normally available under the compliance regime and whether or not the general procedural requirements dictated by Articles 65-68 of the Vienna Convention should apply in such a case.[8]

Finally, the violation of the Kyoto Protocol provisions may also in certain circumstances trigger the application of the law of State responsibility and in particular the taking of countermeasures against the wrongdoer, both in case of a material or of a non-material breach. The applicability of countermeasures may not in fact be ruled out pursuant to the wording of Article 73 of the Vienna Convention, which explicitly states that the provisions of the Vienna Convention do not prejudice any

---

[6] See R. Wolfrum, 'Means of Ensuring Compliance with and Enforcement of International Environmental Law', in *RCADI*, 1998, p. 272.

[7] Art. 60 also refers to the possibility to terminate, and not only to suspend, the operation of a treaty against a Party found to be not in compliance with its obligations, but such an issue is not examined here, since this is not deemed to be really relevant in the context of the Kyoto Protocol on climate change. See also M. Fitzmaurice, cited above, p. 308.

[8] See M. Fitzmaurice, cited above, p. 310.

question that may arise from the international responsibility of a State.[9] In this sense, one may also consider the findings of the International Court of Justice in the *Gabčikovo-Nagymaros* case.[10]

However, with regard to MEAs, in general, and the Kyoto Protocol, in particular, it should not be forgotten that the application of the consequences foreseen by the compliance regime itself embodies important features of the countermeasures, such as the fact that they are taken in response to an unlawful act and their aim is to induce the State to return to compliance. Furthermore, they should be both proportionate and reversible. This could lead to the conclusion that the application of countermeasures, outside and beyond the consequences provided by the compliance regime, should not be allowed under the Kyoto Protocol, insofar as the compliance regime may be said to constitute a sort of *lex specialis* as compared to the general law on countermeasures. However, I am convinced that despite the similarities between the countermeasures and the consequences foreseen by the compliance regime established under the Kyoto Protocol, the possibility to apply countermeasures under the general rules of international law cannot be completely ruled out in this case, although in concrete terms the characteristics of the consequences to be applied within the framework of the compliance regime may well limit the opportunity of having recourse to countermeasures outside the realm of the compliance regime at stake.

2. THE COMPLIANCE REGIME OF THE KYOTO PROTOCOL: SOME BASIC LEGAL ISSUES

Following the remarks on the relationship between the compliance regimes developed under the MEAs and the relevant international law provisions contained in the previous section, it is now time to address some basic legal issues especially related to the compliance regime established by the Kyoto Protocol to tackle disputes among Parties which may arise from its implementation and application.[11]

In this sense, there are at least a couple of paramount preliminary issues which are highly relevant to our analysis and which still await a clear and definitive solution under international law and therefore deserve analysis. The first refers to the

---

[9] On such an issue see also International Law Commission (ILC), Responsibility of States for Internationally Wrongful Acts, 2001, Art. 49 ff.

[10] See ICJ, *Gabčikovo-Nagymaros* case (1997), para. 47: 'Thus the Vienna Convention of 1969 on the Law of Treaties confines itself to defining – in a limitative manner – the conditions in which a treaty may lawfully be denounced or suspended; while the effects of a denunciation or suspension seen as not meeting those conditions are, on the contrary, expressly excluded from the scope of the Convention by operation of Article 73. It is moreover well established that, when a State has committed an internationally wrongful act, its international responsibility is likely to be involved whatever the nature of the obligation it has failed to respect.'

[11] See J. Brunée, 'The Kyoto Protocol: Testing Ground for Compliance Theories?', *ZaöRV*, 2003, p. 255 ff.; F. Yamin & J. Depledge, *The International Climate Change Regime*, Cambridge, 2004, p. 378 ff.; M. Doelle, *From Hot Air to Action*, Toronto, 2005, 109 ff.

legal basis for the approval of the compliance regime developed under the Kyoto Protocol, also known as the 'amendment dilemma'. The second pertains to the determination of the exact relationship of such a regime with the traditional dispute settlement means and systems normally available under international law and recalls the analysis provided in more general terms in the previous section.

The analysis of the first issue ought to start from the provision of Article 18 of the Kyoto Protocol, which directly deals with the issue of non-compliance by Parties with their obligations under the treaty, by stating that:

'The Conference of the Parties serving as the meeting of the Parties to this Protocol shall, at its first session, approve appropriate and effective procedures and mechanisms to determine and to address cases of non-compliance with the provisions of this Protocol, including through the development of an indicative list of consequences, taking into account the cause, type, degree and frequency of non-compliance. Any procedures and mechanisms under this Article entailing binding consequences shall be adopted by means of an amendment to this Protocol.'

As one can see, this provision explicitly calls on the Conference of the Parties (COP) of the Kyoto Protocol to approve, at the first session after the entry into force of the Kyoto Protocol, adequate rules and procedures to tackle potential cases of non-compliance. This, in fact, has regularly occurred in the sense that the KP compliance regime (which had been developed by an *ad hoc* group of experts before the entry into force of the KP and had already been approved on a provisional basis by the seventh session of the Conference of the Parties to the UNFCCC (COP-7) in Marrakech[12]) was officially approved by the Conference of the Parties serving as the Meeting of the Parties (COP/MOP), at its first session (COP/MOP-1), held in 2006 in Montreal.[13]

In this respect, it should be noted that by reading carefully the second part of Article 18, as reproduced above, it can readily be noted that the competence of the COP/MOP is limited to the approval of appropriate 'procedures and mechanisms' to be referred to by the Parties in order to promote compliance with the Kyoto Protocol. However, the wording of Article 18 also makes clear that in case such 'procedures and mechanisms' entail 'binding consequences' for the Parties it is necessary to proceed to their adoption by means of a formal amendment to the Protocol, which is regulated by Article 20 of the KP. Such a provision, in particular, foresees formal 'acceptance' by at least three-fourths of the Parties to the Protocol, which has to occur on the basis of their respective Constitutional provisions and which would normally require a vote by the competent national Parliament, before an amendment duly approved by the COP/MOP may enter into force and demonstrate its full legal effects.

With regard to the Kyoto Protocol and the compliance procedure therein, no formal amendment to the Protocol has so far been either proposed or approved. For

---

[12] See COP7 Decision 24/CP.7 (2001).
[13] See COP/MOP 1 Decision 27/CMP.1 (2005).

this reason, in particular, the so-called 'amendment dilemma' has emerged. Such a 'dilemma' essentially refers to the fact that, when the compliance regime was approved by the COP/MOP1, none of the Parties, with the only limited exception of Saudi Arabia, officially asked for an official amendment procedure under Article 20 of the Kyoto Protocol. This might have been the case because most of the Parties maintained that the dispute avoidance nature of the compliance regime being proposed and approved by the COP/MOP did not envisage within its provisions the presence of 'procedures and mechanisms' entailing binding consequences for the Parties within the meaning of Article 18 of the Kyoto Protocol. Therefore, the COP/MOP decided to proceed with the adoption of the 'procedures and mechanisms' relating to compliance under the KP, without addressing and solving the 'dilemma' related to whether or not it was necessary to approve them through a 'formal' amendment to the Protocol to be ratified by three-fourths of the Parties.

However, if one looks carefully at the consequences that one of the two branches of the Compliance Committee, namely the Enforcement Branch, may apply under the compliance regime adopted by the COP/MOP1, it clearly emerges that there are at least two types of applicable consequences which may hardly be regarded as not falling within the concept of 'procedures and mechanisms entailing binding consequences' for the Parties. These two types of consequences consist in particular of the possibility to proceed to the 'deduction from the Party's assigned amount for the second commitment period of a number of tonnes equal to 1.3 times the amount in tonnes of excess emissions' and in the possibility to declare the 'suspension of the eligibility to make transfers under Article 17 of the Kyoto Protocol', that is from the possibility to make use of the 'international emissions trading' scheme, for a Party which has exceeded its assigned amount under Annex B of the KP.[14]

In both cases, it appears evident that, even without a detailed analysis of the scope and reach of the measures which may be applied by the Enforcement Branch against a non-complying Party, these types of consequences amount to real 'sanctions', which may certainly entail binding consequences for the affected Party. Therefore, it seems that it can hardly be maintained that, at least with respect to the applicability of such consequences, a formal approval of the compliance regime in the form of an amendment to the Kyoto Protocol is not needed.[15]

This notwithstanding, not only was the amendment procedure not commenced, but also the States' practice subsequent to the approval of the compliance regime by the COP/MOP1 seems to point in the opposite direction and contradicts the text of the Kyoto Protocol itself.

In fact, after the approval of the Kyoto Protocol compliance regime by the COP/MOP at its first session in 2005, all the Parties have cooperated in the selection of the members of the Compliance Committee and the new institution has already started its operations. Moreover, the COP/MOP at its second session (2006) consid-

---

[14] See Art. XV, Procedures and Mechanisms relating to compliance under the Kyoto Protocol.

[15] In this sense, see also the explicit reference to Art. 20 KP, besides and along with Art. 18 KP, contained in the Preamble to the 'procedures and mechanisms' adopted by the COP/MOP for the Kyoto Protocol.

ered the first annual report presented by the Compliance Committee on its activities and at the same time adopted the rules of procedure of the Compliance Committee,[16] despite the fact that the 'amendment dilemma' is still unresolved and has not yet been officially addressed by the COP/MOP.

Obviously, State practice which has so far demonstrated the acceptance of the Compliance Committee and its work may not necessarily mean that the issue of the 'amendment dilemma' can be definitively set aside. In fact, most Parties may decide not to contest the legitimacy of the compliance regime as long as the most relevant 'consequences' for a failure to comply with the obligations stemming from the Kyoto Protocol are not really applied by the Enforcement Branch and as long as the activities of the Compliance Committee mainly focus on facilitative aspects, through the work of the Facilitative Branch, as will most likely happen in the next few years. In any case, it is very well possible, and in my personal opinion highly probable, that the 'amendment issue' might arise once again in the future, should the Enforcement Branch eventually decide to make use of the harshest sanctions it has the power to apply under the relevant provisions of the KP compliance regime.

In addition to what I have been saying so far, a second paramount preliminary issue which pertains to the KP compliance regime should be addressed here. It is the issue relating to the determination of the relationship between such a regime and the traditional dispute settlement means and procedures existing under international law. In this respect, in strictly legal terms, the starting point for the analysis ought to be Article 19 of the Kyoto Protocol, which states that:

'The provisions of Article 14 of the Convention on settlement of disputes shall apply mutatis mutandis to this Protocol.'

This explicit referral to the provisions of the UNFCCC contained in the KP is very relevant insofar as it shows the intention of the drafters of the KP not to depart from the basic solutions which had previously been developed within the framework of the UNFCCC, prior to the signature of the KP. Moreover, it is very important since it indirectly enables the creation of a specific link between the operation of the compliance regime especially negotiated for the KP and the possibility to have recourse to the traditional dispute settlement means and procedures existing under international law. Article 14 of the UNFCCC, in this sense, leaves open to the Parties a wide array of options to deal with their disputes, stemming from 'negotiations or any other peaceful means of their own choice' to 'conciliation', without excluding the possibility for the Parties to make a voluntary unilateral declaration amounting to the prior acceptance of the jurisdiction of the International Court of Justice (ICJ) or of the arbitration procedure.

In any case, however, the reference made by Article 19 of the Kyoto Protocol, through Article 14 of the UNFCCC, to the possibility to use the means for a peace-

---

[16] On the activities of the Compliance Committee see the UNFCCC website <www.unfccc.int>. See also the First Report submitted by the Compliance Committee to the COP/MOP (FCCC/KP/CMP/2006/6), approved by the COP/MOP by means of Decision 4/CMP.2.

ful settlement of disputes available under general international law is very important insofar as it clarifies, whether this is necessary or not, that the compliance regime established under the Kyoto Protocol should by no means be considered as an 'exclusive' instrument to deal with disputes arising under the KP.

3. THE COMPLIANCE COMMITTEE AND THE CONSEQUENCES FOR NON-COMPLIANCE

We have been focusing in the previous section on some paramount preliminary legal issues which had to be addressed before proceeding with the analysis of the specific provisions contained in the compliance regime developed under the Kyoto Protocol to prevent and/or settle disputes among the Parties. The time has now come to intensify our analysis of the KP compliance regime. However, it must be stated that it is not the aim of the present contribution to describe once again all the main features of the already well-known compliance regime at stake.[17] Instead, I rather intend to comment on some specific legal issues relating to such a regime, regarding in particular the legal nature, the automatic application and the appeal system envisaged by the KP compliance regime with respect to the consequences which may be applied by the Compliance Committee, through the Facilitative and Enforcement Branches.

The issue of the legal nature of the consequences for compliance which may be applied by the Enforcement Branch has already been touched upon very briefly in the previous section when dealing with the issue of the 'amendment dilemma'. Now I will try to develop the analysis in broader and more comprehensive terms, distinguishing between the measures which may be applied by the Facilitative Branch from those which may be adopted by the Enforcement Branch.

The Facilitative Branch, pursuant to Article IV of the 'Procedures and Mechanisms relating to compliance under the Kyoto Protocol' (hereinafter simply the 'KP Procedures') is responsible for addressing the questions of implementation with the KP by providing advice to the Parties in implementing the Protocol and promoting Parties' compliance with their commitments under the KP, taking into account the principle of common but differentiated responsibilities, as enshrined in Article 3 of the UNFCCC.

The soft and facilitative approach envisaged by the drafters of the compliance regime for the Facilitative Branch, which results from Article IV of the KP Procedures, is also confirmed by the analysis of the types of consequences that it may apply pursuant to Article XIV of the KP Procedures. Such consequences in fact range from advice and the provision of assistance to the Parties on the implementation of the KP provisions, to the facilitation of financial and technical assistance to any of the Parties concerned, and to the formulation of recommendations to any

---

[17] On this issue see, for instance, M. Montini, cited above, p. 175 ff. and F. Yamin & J. Depledge, cited above, p. 378 ff.

Party concerned regarding the promotion of the provision of financial and technical assistance to the other Parties within the scope of the UNFCCC. Needless to say, the analysis of the scope and reach of such measures envisaged by the compliance regime for the Facilitative Branch makes it possible to determine that such provisions do not intend to create a system of binding consequences for non-complying Parties.

The situation seems to be (at least partially) different once the analysis shifts to the activities performed by the Enforcement Branch. In this sense, Article V of the KP Procedures states that this Branch is responsible for three main duties: (i) determining whether a Party included in Annex I is not in compliance with its quantified emission limitation or reduction commitments; (ii) assessing whether the methodological and reporting requirements are respected; and (iii) checking whether the eligibility requirements for all the three flexibility mechanisms foreseen by the KP are fulfilled. All of these three situations specifically relate to the behavior of Annex I Parties and somehow refer to the (non-)fulfillment of their limitation or reduction commitments under the Protocol.

As to the consequences for non-compliance which may be applied by the Enforcement Branch, pursuant to Article XV of the KP Procedures, these may include a declaration of non-compliance and the possibility to request the Party not in compliance to develop a plan containing an analysis of the specific reasons for the non-compliance, an indication of the measures the Party should take to try and restore compliance and the provision of a timetable for the implementation of such proposed measures.

In addition to that, however, Article XV of the KP Procedures also foresees some more specific measures which may be addressed to non-complying Parties under certain conditions. In particular, in this respect one may recall that, on the one hand, when the Enforcement Branch determines that a Party does not fulfill one or more of the eligibility requirements foreseen for the functioning of the flexibility mechanism under Articles 6, 12 and 17 of the Kyoto Protocol, it may suspend the eligibility of that Party to use those mechanisms. Moreover, on the other hand, when the Enforcement Branch determines that a Party has exceeded its assigned amount as listed in Annex B of the KP it may issue a declaration of non-compliance accompanied by three more specific and harsher consequences, namely: (i) the 'deduction from the Party's assigned amount for the second commitment period of a number of tonnes equal to 1.3 times the amount in tonnes of excess emissions', (ii) the 'development of a compliance plan' on more detailed and specific terms than the one already mentioned above with regard to the consequences which may be applied by the Enforcement Branch, and (iii) the 'suspension of the eligibility to make transfers under Article 17 of the Kyoto Protocol', namely under international emissions trading.

The specific measures just mentioned largely seem to depart from the traditional 'facilitative' and 'dispute avoidance' approach which is common to the compliance regimes developed under most MEAs, such as, for instance, with regard to the Montreal Protocol on the protection of the ozone layer, and upon which the Kyoto

Protocol compliance regime is also based. Therefore, I am persuaded that they should be considered as real 'sanctions', rather than facilitative measures foreseen by the compliance regime in order to tackle the failure of a Party to comply with its obligations. As a consequence, I am convinced that they should have been approved by means of a formal amendment to the Kyoto Protocol, as already stated above.

In contrast to this interpretation, however, one may argue that, according to the KP Procedures, not only the consequences applied by the Facilitative Branch, but also those applied by the Enforcement Branch 'shall be aimed at the restoration of compliance to ensure environmental integrity, and shall provide for an incentive to comply.'[18] However, it seems to me that not only the nature but also the specific aim of the measures at stake clearly point in a certain direction. In fact, with regard to the consequences at stake it really seems that a specific 'punitive aim' prevails over the traditional 'restorative aim', which instead largely characterizes the dispute avoidance regimes elaborated under the existing MEAs and, in general terms, also the Kyoto Protocol compliance regime. Therefore, I must conclude that in such a case one cannot refrain from speaking of 'real sanctions' rather than facilitative measures.

Finally, such a conclusion still holds true even if one considers two more objections that could be leveled against the interpretation proposed. The first one is the consideration that such sanctions, despite their appearance, could still be considered as 'soft' measures, insofar as they remain substantially 'internal' to the Kyoto Protocol legal system, in the sense that they are aimed at rendering more 'costly' the fulfillment of its obligation for a non-complying Party, but still presuppose its continued full participation in the operation of such a legal system and do not suspend any rights or privileges other than those created by the Kyoto Protocol itself. The second is the finding that the KP Procedures do not foresee any discretionary power for the Enforcement Branch with regard to the determination of 'the cause, type, degree and frequency of the non-compliance' of the Party against which the harshest measures seen above may be taken,[19] whereas, for instance, a greater discretionary power is foreseen for the less invasive consequences available to the Enforcement Branch. However, in my opinion, none of these two further objections may run counter to the fact that the nature and aim of the specific measures is essentially a 'punitive' rather than a 'facilitative' one, as already noted above.

A different, but still highly interesting feature of the Kyoto Protocol compliance regime which deserves some comment in the framework of the present analysis is represented by the issue of the automatic application of the consequences for non-compliance. In this respect, it must be recalled here that, according to the KP compliance regime, neither the triggering of the compliance mechanisms nor the application of the related consequences by the Facilitative Branch or the Enforcement Branch are dependent on the Parties' consent. In other words, this means that

---

[18] See KP Procedures, Art. V(6).
[19] See KP Procedures, Art. XV(1) and (5). On this issue, see G. Ulfstein, J. Werksman, *The Kyoto Compliance System: Towards Hard Enforcement*; O. Stokke, J. Hovi, G. Ulfstein (eds), *Implementing the Climate Regime. International Compliance*, London, 2005, p. 39.

no Party to the Protocol may escape an investigation and a consequent decision by the Compliance Committee on its enforcement record with regard to the obligations stemming from the Kyoto Protocol provisions.

In fact, this may be considered to be a peculiarity of some of the compliance regimes developed under the existing MEAs, and in particular of the present compliance regime, which largely departs in this respect from the traditional dispute settlement procedures, whose triggering and often the application of whose conclusions is normally still subject to the consent of the Parties concerned, as is generally true for the provisions of any dispute settlement system under international law. The feature of the automatic application of the consequences which characterizes the Kyoto Protocol compliance regime is therefore to be welcomed insofar as it represents an important advancement compared to the operation of traditional dispute settlement procedures. Moreover, such an innovation is particularly relevant in this case, since it may also extend to the application of some specific consequences, which may well be considered as real sanctions, as already explained above.

Finally, the appeal system envisaged by the KP Procedures specifically regarding the consequences applied by the Enforcement Branch of the KP Compliance Committee is worth a few specific comments. According to Article XI of the KP Procedures, a Party in respect to which a final decision has been taken by the Enforcement Branch may decide to appeal to the COP/MOP against such a decision, within 45 days from its notice, if it believes that it has been denied due process. The COP/MOP shall consider the appeal at the first possible session and with a three-fourths majority vote it may decide to override the decision originally adopted by the Enforcement Branch and refer the matter back to it.

Obviously, this appeal procedure creates some doubts as to its legitimacy to the purist of the dispute settlement and compliance regimes, insofar as it introduces an element of a 'political' nature within a regime otherwise characterized by an essentially 'legal' nature. The doubts obviously arise since the COP/MOP is essentially a political body composed of representatives of Party Governments. This is certainly true, but a final word on the real scope and reach of such a feature should in my opinion be suspended until the first appeals are decided by the COP/MOP as we still do not know how they will be dealt with. One should in fact consider that in other international treaties where a similar feature is present, such as, for instance, in the WTO Dispute Settlement Regime (which, however, is characterized by the so-called 'inverted consensus' procedure) the practical relevance of the (potential) political influence of the Dispute Settlement Body (which is essentially a political body composed of Party representatives) on the legal decisions by the Panel and the Appellate Body has so far been absent or totally negligible.

4. THE RESPONSIBILITY FOR NON-COMPLIANCE UNDER INTERNATIONAL
    LAW AND EC LAW COMPARED

A very interesting issue raised by the Kyoto Protocol compliance regime is represented by the complex and very appealing issue of responsibility for non-compliance, particularly in the case where some Parties proceed to the joint fulfillment of their obligations under the Protocol, pursuant to Article 4 of the KP. In such a case, in fact, the issue of the responsibility of a Party for non-compliance may present different features depending on whether it is addressed from an International law or from an EC law perspective, as I will try to highlight.[20]

In this respect, Article 4 of the Kyoto Protocol foresees the possibility for 'any Parties included in Annex I' of the UNFCCC to agree on the joint fulfillment of their obligations under the Kyoto Protocol, either outside or within the framework of a regional economic organization. This provision has obviously been created mainly to accommodate the needs of the European Community, which is so far the only non-State Party to the Kyoto Protocol together with its member states, although it does not exclude the possibility for other Parties, which are not acting within the framework of a regional economic organization, to proceed to the joint fulfillment of their obligations stemming from the Kyoto Protocol.

In this sense, it should be recalled, first of all, that once some of the Parties to the Kyoto Protocol have reached an agreement on the joint fulfillment of their obligations and have duly notified this to the UNFCCC Secretariat, they become bound by their respective obligations laid down in the subsequent agreement, which essentially modifies their initial commitment originally laid down in Annex B to the Kyoto Protocol. However, as already mentioned above, the practical features of their respective responsibility essentially depend on whether or not the Parties concerned, which have reached an agreement on the joint fulfillment of their obligations, are members of a regional economic organization.

In fact, with respect to Parties acting outside the framework of a regional economic organization, Article 4(5) of the KP states that:

'In the event of failure by the Parties to such an agreement to achieve their total combined level of emission reductions, each Party to that agreement shall be responsible for its own level of emissions set out in the agreement'

while in the case of Parties which belong to and act within the framework of a regional economic organization, such as the European Community, Article 4(6) of the KP foresees that:

'If Parties acting jointly do so in the framework of, and together with, a regional economic integration organization which is itself a Party to this Protocol, each member State of that regional economic integration organization individually, and together with the regional economic integration organization acting in accordance with Article 24, shall, in the event of failure to achieve the total combined level of emission reductions, be responsible for its level of emissions as notified in accordance with this Article'.

---

[20] On the EC law perspective see F. Jacquemont, 'The Kyoto Compliance Regime, the European Bubble: Some Legal Consequences', in M. Bothe & E. Rehbinder (eds.), *Climate Change Policy*, Utrecht, 2005, p. 352 ff.

The two situations essentially differ depending on the involvement of the only regional economic organization which is so far a Party to the Kyoto Protocol, namely the European Community. In fact, if one or more member state(s) is (are) not fulfilling its (their) obligations to the extent that its (their) failure also implies the failure of the EC to meet its overall collective target, the responsibility for the failure to fulfill their commitments will extend to both the defaulting State(s) and the European Community itself.

With respect to such an issue, several legal questions may arise. The first and foremost question relates to the issue of the respective responsibility of the Parties to the joint fulfillment agreement. By reading the text of Article 4 of the KP it cannot in fact be easily determined whether in the case of such a failure to fulfill their respective commitments both the EC and all its member states will be responsible under international law or whether, instead, responsibility will arise merely for the EC together with the defaulting State(s).

The wording of the KP, in this respect, is not completely clear, and the question can only be solved by referring to the 'context' and to the 'object and purpose' of the Kyoto Protocol, rather than sticking to the plain textual data, as mandated in such cases by the general rule on the interpretation of international treaties contained in Article 31 of the Vienna Convention on the Law of Treaties. In this context, in fact, if one places the provision of Article 4 of the KP within the framework of the 'context' and the 'object and purpose' of the Kyoto Protocol, it emerges that the most logical interpretation can only be the one which makes just the defaulting State(s) responsible towards the other Parties, together with the regional economic organization, namely the EC, rather than all the EC member states which are Parties to the agreement on the joint fulfillment of their obligations.

Moreover, the second highly relevant question which arises here pertains to the additional responsibility for the failure to fulfill their commitments which may arise under EC law from the failure of one or more of the EC member states to satisfy their obligations. In such a case, in fact, it is to be understood that besides the responsibility stemming under International law for the defaulting State(s) (and possibly for the EC itself), there are some additional consequences which may arise under EC law for the relevant State(s) for the failure to fulfill its (their) commitments under the Kyoto Protocol.

The main reason for that lies in the fact that once the modified commitments undertaken by the EC member states within the framework of the agreement on the joint fulfillment of their obligations, commonly indicated as the 'EC Burden Sharing Agreement', were agreed upon by the Environment Ministers of the EC member states in 1998, they were then transferred into a binding Decision by the EC Council in 2002.[21]

This may obviously entail very relevant consequences under EC law if some of the commitments agreed upon are not fulfilled by one or more of the interested

---

[21] See EC Council Conclusions of 16-17 June 1998 (para. 2 and Annex I) and EC Council Decision 2002/358 of 25 April 2002 (Annex II) concerning the approval on behalf of the EC of the Kyoto Protocol.

Parties.[22] In such a case, in fact, a Party to the EC Burden Sharing Agreement which does not fulfill its obligations under the agreement may be sanctioned under EC law on the basis of the combined application of Article 10 of the EC Treaty, which contains the general loyalty principle that applies to all the relationships of the EC member states towards the EC legal system and the EC institutions, and of the specific provisions contained in Decision 2002/358/EC by means of which the EC Council has approved the Kyoto Protocol on behalf of the EC.

On the basis of such provisions, it can be assumed that an EC member state which fails to fulfill its obligations under the KP, as supplemented by the EC Burden Sharing Agreement, may be subject to an infringement procedure under Article 226 of the EC Treaty by the European Commission for its violation of the applicable EC law provisions just mentioned above.

This is the situation in purely legal terms. In practical terms, however, it is still unclear what the concrete disadvantage of this double responsibility for a non-complying EC member state could be, both under International law and under EC law. In fact, in concrete terms the effectiveness of the possible sanctions emanating from EC law, which may be much stricter and more punitive in nature and therefore theoretically exercise a greater deterrent role for the States concerned, may be severely limited by the fact that the ascertainment of the position of a defaulting State under EC law with respect to the obligation stemming from the KP and the EC Burden Sharing Agreement would probably take place very late, well after the expiry of the fist commitment period foreseen by the Kyoto Protocol. This is essentially due to the fact that, nowadays, the usual duration of the infringement procedures under Article 226 EC Treaty is normally around two or three years. This is obviously not very efficient, particularly in a sector such as climate change, which is characterized by high costs of compliance for the Parties and which should rather be assessed and planned as early as possible.

---

[22] It should be recalled here that according to Art. 4(4) KP, 'If Parties acting jointly do so in the framework of, and together with, a regional economic integration organization, any alteration in the composition of the organization after adoption of this Protocol shall not affect existing commitments under this Protocol.' As a consequence, the EC Burden Sharing Agreement only relates to the 15 States which where Parties to the EC when the KP was signed.

IMPLEMENTATION OF THE KYOTO PROTOCOL IN GERMANY: DESIGNING AN INTEGRATED MANAGEMENT SCHEME FOR GREENHOUSE GASES

Michael Mehling[*]

1. INTRODUCTION

Seen from the international arena, Germany may come across as the self-declared standard- bearer of a sustainable energy revolution, a tireless campaigner for the paradigmatic transformation of energy practices in industrialized nations. Concurrently holding the presidencies of the Group of Eight (G8) and the Council of the European Union (EU), it has already indicated its intention to promote an ambitious agenda on energy and climate issues during the first half of 2007.[1] And as a regular advocate of stringent measures during climate negotiations, the official host to the Secretariat of the United Nations Framework Convention on Climate Change (UNFCCC),[2] and a market leader in the promotion and deployment of progressive energy technologies, it is commonly perceived as an influential actor in international climate policy.

However, Germany is not only attracting praise for its efforts abroad; its domestic policies on the production, conversion, distribution and utilization of energy – which have featured an environmental dimension for nearly two decades[3] – are equally acclaimed for their success in reducing greenhouse gas emissions and declared a prototype for legislation in other countries.[4] More recently, however, and

---

[*] Assessor iur., LL.M.; Research Fellow, Faculty of Law, University of Greifswald, Germany; Associate, Ecologic – Institute for International and European Environmental Policy.

[1] For further information on the German work programme during its presidencies of the EU Council and the G8, see Federal Government, *Europe – Succeeding Together: Presidency Programme 1 January to 30 June 2007*, available on the Internet at: <www.eu2007.de> (last accessed on 15 April 2007), 15; the website of the German G8 presidency at: <www.g-8.de> (last accessed on 15 April 2007); and the recommendations for the presidency by the German Advisory Council on Global Change (WBGU), *New Impetus for Climate Policy: Making the Most of Germany's Dual Presidency*, available on the Internet at: < www.wbgu.de/wbgu_pp2007_engl.pdf> (last accessed on 15 April 2007).

[2] The Secretariat is located in Bonn, see the treaty between Germany, the United Nations and the Convention Secretariat, Federal Law Gazette (*BGBl.*) Part II (1996), p. 2782.

[3] See already Jeannine Cavender and Jill Jäger, 'The History of Germany's Response to Climate Change,' 5 *International Environmental Affairs* (1993), pp. 3 et seq.

[4] In its widely-read annual report, for instance, the Worldwatch Institute drew attention to German legislation on renewable energy sources as a successful model, see Janet L. Sawin, 'Charting a

W.Th. Douma et al., eds., The Kyoto Protocol and Beyond
© 2007, T·M·C·ASSER PRESS, *The Hague, The Netherlands and the Authors*

largely in response to legislative impulses from the European Community, the regulation of greenhouse gases has shifted away from a segmented array of isolated measures and initiatives on specific aspects of global warming, such as policies to manage energy demand or to promote research on sustainable alternatives, to an increasingly sophisticated network of regulatory standards, market mechanisms, and other innovative approaches.

While the first elements of a new area of law are arguably emerging in the shape of common principles and objectives for sustainable energy use,[5] the countless rules devoted to climate change are still only loosely related and far from becoming a coherent normative framework. Partly, this can be ascribed to a dramatic change in the conception and focus of environmental regulation: as economic considerations acquire greater weight in decision making, an increased preoccupation with the cost and efficiency of policies has resulted in a variety of flexible market incentives joining or supplanting more conventional performance and quality standards.[6]

With energy production and consumption accounting for the vast majority of anthropogenic greenhouse gas emissions, moreover, climate policy invariably affects larger and also more sensitive areas of society, compelling change in nearly all domains of social behavior and, notably, constraining economic activity at a much broader scale than any other area of environmental governance. As a result, decision makers have openly embraced alternative policy approaches based on flexible markets and price incentives, in the hope of limiting harmful effects on the economy and competitive distortions in the global marketplace. While the reasoning behind this changed orientation is understandable, the rapid growth in and the evolution of new mechanisms has also brought along new shortcomings, giving rise to conflicts at the level of individual rules and principles, all the way to systemic tensions within the overall configuration of the legal system.

Such difficulties have also overshadowed the implementation of the Kyoto Protocol[7] in Germany. Looking back at the early stages of this process, one might garner the impression of an incremental, barely coordinated strategy, resulting in a coincidental rather than intended assortment of regulatory devices, not seldom based on overly rushed legislative schedules,[8] substantive disagreement between rival

---

New Energy Future' in: Gary T. Gardner and Linda Starke, *State of the World 2003* (New York: W.W. Norton, 2003), pp. 85 et seq.

[5] On this development, see Michael Rodi, 'Grundstrukturen des Energieumweltrechts', 3 *Europäisches Umwelt- und Planungsrecht* (2005), pp. 165 et seq.

[6] See fundamentally Tom Tietenberg, 'Economic Instruments for Environmental Regulation', 6.1 *Oxford Review of Economic Policy* (1990), at pp. 17 et seq.

[7] Kyoto Protocol to the United Nations Framework Convention on Climate Change, Kyoto, 10 December 1997, in force 16 February 2005, U.N. Doc. FCCC/CP/1997/L.7/Add.1, (1998) 37 *International Legal Materials*, 22.

[8] One might also draw attention to the current approach to political representation, which favors short-term measures over long-term strategic policies by exerting pressure on elected politicians to provide demonstrable results in time for the next popular vote, see generally Anthony Downs, *An Economic Theory of Democracy* (New York, N.Y.: Harper & Row, 1957); Joseph A. Schumpeter, *Capitalism, Socialism and Democracy* (New York, N.Y.: Harper & Row, 1942).

government agencies, and the challenge of balancing international commitments with domestic legal and political realities. Faced with changing demands in a politically exposed issue area, legislators and administrators have been mandated with elaborating an operational regime for activities which, previously, had been subject to no form of regulation. Confused by the unfolding disarray and widespread misinformation, industrial operators and other stakeholders have understandably voiced their irritation at the lack of coherence and systematization in German climate policy.

And yet, as this area of law matures, one can already perceive efforts to streamline the current diversity of rules through shared definitions, common objectives, and dynamic referencing between different acts of legislation. Against the backdrop of a major initiative to systematize the diversity of environmental statutes, ordinances, decrees, and other sources of environmental law in a uniform code,[9] it should hardly come as a surprise that suggestions have also been made to harmonize German climate policy under a single umbrella act, marking a departure from piecemeal regulation to an integrated system for the management of our atmosphere. Based on an outline of current legislative efforts to implement the Kyoto Protocol, this chapter will seek to identify the potential for such an overarching management regime and its central implications for the operation of future climate policies in Germany. Implicitly, it also asks whether such an approach, once deployed in Germany, might again serve as a model for domestic climate policies in other states faced with similar challenges as they manoeuvre in a sophisticated, polycentric legal system.

2.   CLIMATE POLICY IN GERMANY: A MODEL OF DIVERSITY

2.1   **Ambitious objectives – A myriad of solutions**

In global climate negotiations, the European Union has consistently supported the adoption of quantified commitments on greenhouse gas reduction. Unsurprisingly, its member states were also among the first industrial nations to ratify the Kyoto Protocol, binding themselves to an average reduction of 8% relative to 1990 levels by 2012. This reduction target has been distributed within the European Union through an elaborate burden-sharing agreement,[10] or 'bubble', under which the Federal Republic of Germany pledged to reduce its greenhouse gas emissions by

---

[9] On this codification process, see Eberhard Bohne, 'Das Umweltgesetzbuch vor dem Hintergrund der Föderalismusreform', 4 *Europäisches Umwelt- und Planungsrecht* (2006), pp. 276 et seq.

[10] See Annex II of Council Decision 2002/358/EC of 25 April 2002 concerning the approval, on behalf of the European Community, of the Kyoto Protocol to the United Nations Framework Convention on Climate Change and the joint fulfillment of commitments thereunder, OJ L 130, pp. 1 et seq.; Germany finalized the domestic ratification process on 26 April 2002 with the adoption by the Federal Council, or *Bundesrat*, see BGBl. Part II (2002), pp. 966 et seq.

21% by 2012. While it is not even certain that Germany will meet this international target, it has since committed itself to still more stringent reductions with a domestic Climate Protection Programme adopted in July 2005.[11]

After the European Council stated its 'firm independent commitment to achieve at least a 20 per cent reduction of greenhouse gas emissions by 2020 compared to 1990' in March 2007,[12] Sigmar Gabriel, the Federal Minister for the Environment, Nature Conservation and Nuclear Safety, outlined a set of concrete measures to achieve this objective in Germany.[13] Among the measures he proposed to the Federal Parliament in April 2007 were:

- a reduction of electricity consumption by 11% through major improvements in energy efficiency;
- an overhaul of the power plant fleet by the replacement of older plants with more efficient ones;
- increasing the share of renewable energy in electricity generation to more than 27%;
- doubling the efficient use of combined heat and power generation to 25%;
- reducing energy consumption by the modernization of buildings, efficient heating facilities and production processes;
- increasing the share of renewable energy in the heating sector to 14%;
- increasing efficiency in the traffic sector and expanding the share of biofuels to 17%; and
- reducing emissions of other greenhouse gases, such as methane, by 40 million tonnes of CO_{2eq}.

The Climate Protection Programme of 2005 details a broad range of policies designed to meet the foregoing commitments. Specifically, this strategy has brought forth legislation on the promotion of renewable energy, guaranteeing prices for electricity obtained from renewable sources and binding grid operators to purchase it,[14] and acceptance and remuneration provisions for electricity generated through

---

[11] See Federal Government *Nationales Klimaschutzprogramm 2005*, Decision of 18 July 2005, Federal Records of Parliament (*BT-Drs.*) 15/5931, at p. 50, where the federal government has pledged to reduce its own CO_2 emissions by 40% by 2020 relative to 1990 levels, provided the member states of the European Union collectively pledge to achieve a 30% reduction in the same period; this programme expands on an earlier programme of October 2000.

[12] Council of the European Union, Brussels, 8 and 9 March 2007, Presidency Conclusions, para. 32 et seq., available on the Internet at: <http://www.consilium.europa.eu/uedocs/cms_Data/docs/pressdata/en/ec/93135.pdf> (last accessed on 15 April 2007).

[13] Sigmar Gabriel, *Klimapolitik der Bundesregierung nach den Beschlüssen des Europäischen Rates Klimaagenda 2020: Regierungserklärung vor dem Deutschen Bundestag*, Berlin, 26 April 2007, available on the Internet at: <www.bmu.de/reden/bundesumweltminister_sigmar_gabriel/doc/39239.php> (last accessed on 30 April 2007).

[14] See Sections 4 et seq. of the Renewable Energy Sources Act (*Gesetz für den Vorrang Erneuerbarer Energien*) of 21 July 2004, BGBl. Part I (2004), pp. 1918 et seq.

combined heat and power generation.[15] Support for biomass and biofuels in the heating and transport sectors is governed by separate legislation.[16]

The efficient use of energy, in turn, is promoted through various rules focusing on heating facilities and the thermal insulation of new buildings.[17] A labeling scheme involving a declaration of the energy performance of household and office appliances seeks to harness the power of consumers to influence product design through purchasing decisions.[18] Further measures have included financial incentives in the energy, transport, and construction sectors, including subsidies for new photovoltaic installations,[19] support for improved insulation of existing buildings,[20] and initiatives to improve the fuel efficiency of vehicles and impose charges on traffic emissions.[21]

A further cornerstone of the strategy outlined in the Climate Protection Programme is an Ecological Tax Reform, designed to reduce the tax burden on labor and shift part of it to energy consumption. Launched on 1 January 1999, this reform increased tax rates on mineral oil and gas, and introduced a new levy on electricity, prompting an annual rise in energy prices between 1999 and 2003.[22] On 1 August 2006, the German legislator adopted a uniform Energy Tax Act setting out a common fiscal framework for energy products through harmonized definitions, taxation rules, and exemptions.[23] The application of this new act is facilitated by an ordinance guiding the implementation of individual provisions.[24]

---

[15] See Sections 4 and 7 of the Combined Heat and Power Act (*Gesetz für die Erhaltung, die Modernisierung und den Ausbau der Kraft-Wärme-Kopplung*) of 1 April 2002, BGBl. Part I (2002), pp. 1092 et seq.

[16] See the Biomass Ordinance (*Verordnung über die Erzeugung von Strom aus Biomasse*) of 21 June 2001, BGBl. Part I (2001), pp. 1234 et seq., and the Biofuel Quota Act (*Gesetz zur Einführung einer Biokraftstoffquote durch Änderung des Bundes-Immissionsschutzgesetzes und zur Änderung energie- und stromsteuerrechtlicher Vorschriften*) of 18 December 2006, BGBl. Part I (2006), pp. 3180 et seq.

[17] See, in particular, the Energy Saving Act (*Gesetz zur Einsparung von Energie in Gebäuden*) of 1 September 2005, BGBl. Part I (2005), pp. 2684 et seq., and the Energy Saving Ordinance (*Verordnung über energiesparenden Wärmeschutz und energiesparende Anlagentechnik bei Gebäuden*) of 2 December 2004, BGBl. Part I (2004), pp. 3146 et seq.

[18] See the Energy Consumption Labelling Act (*Gesetz zur Umsetzung von Rechtsakten der Europäischen Gemeinschaften auf dem Gebiet der Energieeinsparung bei Geräten und Kraftfahrzeugen*) of 30 January 2002, BGBl. Part I (2002), pp. 570 et seq.

[19] These included, for instance, the Market Launch Programme for Renewable Energy Sources and the 100,000 Roofs Programme, see Federal Ministry for the Environment, Nature Conservation and Nuclear Safety, *Germany's National Climate Protection Programme* (Berlin: BMU, 2000), p. 7.

[20] Climate Protection Programme for Existing Buildings, which provided low-interest loans from the Kreditanstalt für Wiederaufbau (KfW Group).

[21] These have included additional investments in rail infrastructure, a motorway toll for commercial transport, the promotion of fuel-efficient cars through tax exemptions under the motor vehicle tax, agreements with car manufacturers on reduced fuel consumption, information and public education campaigns, as well as landing fees for airports.

[22] For a general overview, see Michael Mehling, 'The Ecological Tax Reform in Germany', 26 *Tax Notes Int'l* (2000), pp. 871 et seq.

[23] Energy Tax Act (*Gesetz zur Neuregelung der Besteuerung von Energieerzeugnissen und zur Änderung des Stromsteuergesetzes*) of 15 July 2006, BGBl. Part I (2006), pp. 1534 et seq.; see Michael Mehling, 'Germany's New Energy Tax Act – A Sign of Progress?', in Jon Almeras (ed.), *Energy: A Special Supplement* (Arlington, VA: Tax Analysts, 2006), pp. 132 et seq.

[24] Energy Tax Implementation Ordinance (*Verordnung zur Durchführung energiesteuerrechtlicher Regelungen und zur Änderung der Stromsteuer-Durchführungsverordnung*) of 31 July 2006, BGBl. Part I (2006), pp. 1753 et seq.

Emissions from industrial facilities, which collectively make up the main source of greenhouse gases in Germany, have conventionally been targeted by a general duty to use energy efficiently under ambient pollution control law.[25] With the adoption of an emission allowance trading system by the European Union,[26] however, operators of a range of specified installations are now faced with a mandatory limitation, or 'cap', on their emissions. Since 1 January 2005, these operators have been required to obtain a permit for CO_2 emissions and must surrender a sufficient number of allowances each year to cover their emissions during the previous year. Germany has adopted federal legislation governing the allocation and distribution of allowances, monitoring of transactions, establishment of the national registry, penalties, and domestic and international reporting procedures.[27]

On a voluntary basis, moreover, German industry has committed itself to reductions in greenhouse gas emissions, replacing an earlier 'Declaration on Global Warming Prevention' with a more ambitious 'Agreement on Climate Protection between the Government of the Federal Republic of Germany and German Business' in November 2000.[28] In exchange for a suspension of further regulatory measures by the Government, major sectors of industry, represented by the leading trade associations, agreed to reduce their emissions of CO_2 and other greenhouse gases by specified amounts before 2012.[29] Monitoring of the performance under this voluntary commitment will occur through an independent scientific institution, with questions on implementation and interpretation left to an advisory committee.[30]

2.2 A wealth of regulation – A dearth of achievement?

According to the first Demonstrable Progress Report compiled under Article 3(2) of the Kyoto Protocol, Germany had implemented in excess of 60 different policies

---

[25] See Section 5(1) Nos. 2 and 4 of the Federal Ambient Pollution Control Act (*Gesetz zum Schutz vor schädlichen Umwelteinwirkungen durch Luftverunreinigungen, Geräusche, Erschütterungen und ähnliche Vorgänge*) of 21 March 1974, in the amended version of 26 September 2002, BGBl. Part I (2002), p. 3830.

[26] See Directive 2003/87/EC of the European Parliament and of the Council of 13 October 2003 establishing a Scheme for Greenhouse Gas Emission Allowance Trading within the Community and Amending Council Directive 96/61, OJ (2003) L275, pp. 32 et seq.

[27] Greenhouse Gas Emissions Trading Act (*Gesetz über den Handel mit Berechtigungen zur Emission von Treibhausgasen*) of 8 July 2004, BGBl. I (2004), pp. 1578 et seq.; further statutory laws deal with the allocation of allowances and the recognition of credits under the Kyoto Protocol, see Allocation Act (*Gesetz über den nationalen Zuteilungsplan für Treibhausgas-Emissionsberechtigungen in der Zuteilungsperiode 2005 bis 2007*) of 26 August 2004, BGBl. Part I (2004), pp. 2211 et seq., and Project Mechanisms Act (*Gesetz über projektbezogene Mechanismen nach dem Protokoll von Kyoto zum Rahmenübereinkommen der Vereinten Nationen über Klimaänderungen vom 11. Dezember 1997*) of 22 September 2005, BGBl. Part I (2005), pp. 2826 et seq.

[28] Agreement on Climate Protection between the Government of the Federal Republic of Germany and Business Federation of German Industries of 9 November 2000, available on the Internet at <www.bmu.de/en/1024/js/topics/climateprotection/agreement> (last accessed on 15 April 2007).

[29] See sections I and II of the Agreement, *supra* n. 28.

[30] See sections IV and VI of the Agreement, *supra* n. 28; responsibility for monitoring has been assigned to an independent institution, the Rheinisch-Westfälische Institut für Wirtschaftsforschung (RWI).

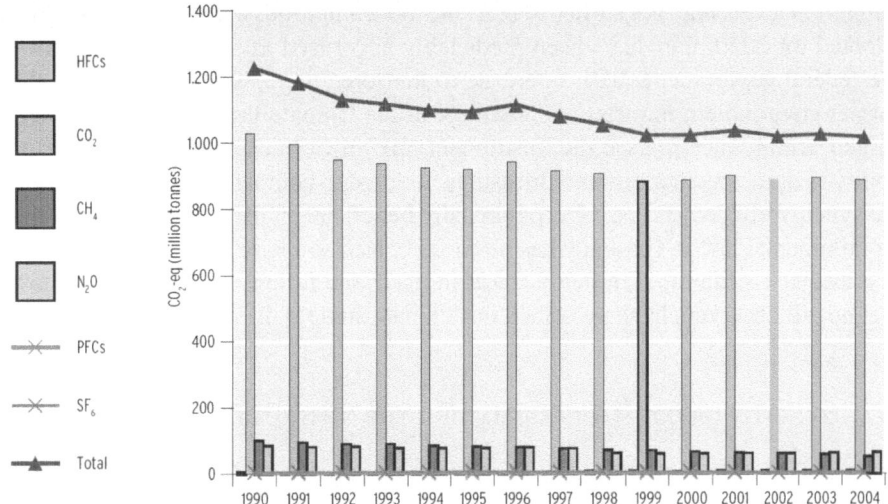

Fig. 1. Development of Greenhouse Gas Emissions in Germany 1990-2004 (*Source:* Federal Ministry for the Environment, Nature Conservation and Nuclear Safety, 2006)

and measures by July 2005.[31] Clearly, then, the German legislator has not hesitated to survey different regulatory approaches and adopt a wide range of policies and measures for its Climate Protection Programme. Given its scope and diversity, however, one is invariably compelled to ask whether the foregoing strategy has been successful in mitigating emissions. Recent data on the greenhouse gas output by German sources imply that emissions in 2004 had fallen by 17.4% relative to the base years of 1990 through 1995, equaling a reduction of 211 million tonnes of CO_{2eq}.[32] A large part of this reduction is, of course, due to the 'wall-fall profits' of German reunification and the virtual collapse of many industries in the former German Democratic Republic (GDR).[33] On the basis of model calculations, however, emissions of CO_2 flowing from the energy sector and industrial sources are forecast to fall to 860 million tonnes in 2010, as opposed to around 936 million tonnes without current policy measures.

Arguably, then, the reduction of 76 million tonnes incurred by political and legal incentives will translate into a 7.5% reduction compared with emission levels in 1990, or roughly half of the overall decline in greenhouse gases.[34] While such

---

[31] Equalling an absolute reduction to 1,015.7 million tonnes CO_{2eq} from 1,226.7 million tonnes CO_{2eq}, see Federal Ministry for the Environment, Nature Conservation and Nuclear Safety, *Demonstrable Progress Report 2006 Report under the United Nations Framework Convention on Climate Change* (Berlin: BMU, 2006), pp. 8 et seq.

[32] Ibid., p. 23.

[33] Joachim Schleich, Wolfgang Eichhammer, Ulla Boede, Frank Gagelmann, Eberhard Jochem, Barbara Schlomann, and Hans-Joachim Ziesing, 'Greenhouse Gas Reductions in Germany: Lucky Strike or Hard Work?', 1 *Climate Policy* (2001), pp. 363 et seq.

[34] Umweltbundesamt, *Klimaschutz in Deutschland bis 2030, Endbericht zum Forschungsvorhaben Politikszenarien III* (Berlin: UBA, 2005).

mitigation levels may not suffice to meet the more ambitious commitments recently adopted for 2020, they have been predictably advertised as a political success by the federal government. With reference to the foregoing basket of measures, the former environment minister and architect of the Climate Protection Programme, Jürgen Trittin, once praised the 'instrument mix' in German climate policy for its ability to draw together different measures 'tailored to complement each other, create synergy effects, and prove economically beneficial'.[35] But while it is true that environmental law in Germany has never embraced so many varied approaches in one thematic strategy, admittedly a feat in itself, the following section will show that not all observers have described this climate strategy in such favorable terms.

3.   SYNERGY OR CONFLICT? DISSECTING THE 'INSTRUMENT MIX'

3.1   Global Warming – A new kid on the block?

Growing in consecutive stages, the body of rules devoted to climate policy has evolved into a comprehensive and highly diverse regulatory strategy. And as was shown in the preceding section, it has proven largely successful in achieving its objective of reducing greenhouse gas emissions. But as with most entities that develop over time, it has not always grown in a systematic fashion, rather adding layer upon layer to accommodate new challenges and international commitments. In recent years, therefore, the German climate strategy has been subject to growing criticism for consisting of 'several barely coordinated measures and actions' whose 'interaction, mutual enhancement, and mutual cancellation' are not fully known.[36] Its instruments have been censored for 'being introduced, modified or expanded in a random manner', resulting in regulatory overlap and excessive government intervention, all of which, in turn, is 'stifling the market'.[37]

While these verdicts mostly originate from representatives from trade and industry, the sectors most affected by environmental and energy policies, they are not

---

[35] Federal Ministry for the Environment, Nature Conservation and Nuclear Safety, *Von Ökosteuer bis Emissionshandel*, Press Release 78/04 of 23 March 2004, quoting the former environment minister, Jürgen Trittin: 'Der Instrumentenmix zum Klimaschutz in Deutschland dient dazu, die Verpflichtung aus dem Kioto-Protokoll zu erfüllen. Die Klimaschutzinstrumente in Deutschland sind maßgeschneidert, ergänzen sich gegenseitig, schaffen Synergieeffekte und lohnen sich volkswirtschaftlich.'

[36] Carsten Kreklau, Commercial Manager of the Federation of German Industries (BDI), in the *Süddeutsche Zeitung* of 17 July 2001, published on the Internet at <www.sueddeutsche.de/deutschland/artikel/162/9153/> (last accessed on 15 April 2007): 'Das gegenwärtige Instrumentarium zur Klimavorsorge besteht bereits aus vielen, kaum aufeinander abgestimmten Maßnahmen und Aktionen. Die Wechselwirkungen, die gegenseitige Verstärkung sowie die Auslöschung zwischen den bereits jetzt bestehenden Instrumenten sind noch nicht in vollem Umfang bekannt. Es geht vor allem (...) um die ungeklärten Wechselwirkungen und daraus resultierenden Begrenzungen wirtschaftlicher Tätigkeit.'

[37] Wirtschaftsrat der CDU e.V., *Macht der Emissionshandel den bestehenden Instrumentenmix überflüssig?* (Berlin: Wirtschaftsrat, 2004).

entirely unfounded: even an Advisory Council of the federal government has observed that interactions between different policies have been 'insufficiently considered',[38] suggesting that the German basket of instruments for greenhouse gas mitigation deserves further attention. Generally speaking, such an instrument mix can be the outcome of a carefully guided process, or merely the accidental convergence of various measures adopted by decision makers in a political system to achieve a set objective.[39]

Leaning more towards the latter category, it appears, global warming legislation has been adopted over time and in response to situational demands, sacrificing systemic coherence for a profusion of divergent terminologies and altogether various degrees of overlap, ambivalence and inconsistency.[40] Important issues are frequently governed by executive ordinances and decrees rather than statutory law, constituting a violation of the constitutional doctrine of essentiality, which requires that substantial issues be governed by formal parliamentary acts.[41]

With energy and environmental regulation in the member states largely initiated by Community law, many of the foregoing shortcomings can be traced back to the supranational level, where the adoption of legislation is a process strongly guided by regulatory competition between the member states[42] and often finds its basis in a precarious compromise in the Council. For the time being, therefore, both the domestic and the Community legislator are accused of excess regulation, undue levels of bureaucracy, and sluggish reform agendas.

Looking back in time, these challenges might also find their origin in the very history of environmental legislation, which evolved from earlier rules on trade supervision and traditionally relied on a rigid system of administrative permits and control.[43] Its ambit was commonly limited to the regulation of impending threats to public safety, such as acute pollution and other perilous activities, rather than dis-

---

[38] Wissenschaftlicher Beirat beim Bundesministerium für Wirtschaft und Arbeit, 'Zur Förderung erneuerbarer Energien', 15 *Zeitschrift für Umweltrecht* (2004), pp. 400 et seq., at p. 401.

[39] Georg Hermes, 'Instrumentenmix im Energieumweltrecht', in Martin Führ, Rainer Wahl and Peter von Wilmowsky (eds.), *Umweltrecht und Umweltwissenschaft: Festschrift für Eckard Rehbinder* (Berlin: Erich Schmidt, 2007), pp. 569 et seq., at p. 572.

[40] Assessing the larger body of environmental law, of which legislation on climate change forms a part, the Federal Ministry for the Environment, Nature Conservation and Nuclear Safety has itself affirmed these shortcomings, see the study by Hans-Heinrich Lindemann, *Denkschrift für ein Umweltgesetzbuch* (Berlin: Erich Schmidt, 1994), at pp. 14 et seq.

[41] This doctrine is derived from the principle of democracy contained in Art. 20(1) of the German Basic Law (*Grundgesetz für die Bundesrepublik Deutschland*) of 23 May 1949, BGBl. Part I (1949), pp. 1 et seq.

[42] See generally Adrienne Héritier, Christoph Knill and Susanne Mingers, *Ringing the Changes in Europe: Regulatory Competition and the Transformation of the State* (Berlin: de Gruyter, 1996), *passim*.

[43] See Gerhard Feldhaus, 'Zur Geschichte des Umweltrechts in Deutschland', in: Klaus-Peter Dolde (ed.), *Umweltrecht im Wandel* (Berlin: Erich Schmidt, 2001), pp. 15 et seq., at pp. 17-9; Klaus-Georg Wey, *Umweltpolitik in Deutschland: Kurze Geschichte des Umweltschutzes in Deutschland seit 1900* (Opladen: Westdeutscher Verlag, 1982), pp. 27 et seq., 105-27, pointing to the origins of modern pollution legislation in the area of 'Gewerberecht' and its close relationship with measures to avert danger, or 'Gefahrenabwehr', still found in current police legislation.

tant, elusive environmental risks.[44] Given their innate affinity to pollution prevention and control, however, measures taken to mitigate global warming were initially assigned to the same area of law governing noise and air pollution. Accordingly, a central act of legislation in this field, the Federal Ambient Pollution Control Act, mentions protection of the atmosphere amongst its objectives,[45] which is commonly understood to include the global climate.[46]

And yet, the very notion of climatic change has, by definition, originated from a precautionary outlook, seeing how it involves diffuse, cumulative manifestations of risk rather than localized and immediate danger. Unlike conventional pollutants, therefore, greenhouse gases were not subject to any form of management in the past, with the ability to emit greenhouse gases only limited by the capacity of an installation.[47] Elaborating climate policies within the regulatory ambit of pollution control is, however, proving less and less viable, as legislators are compelled by economic constraints and supranational commitments to engage in a paradigmatic shift of regulatory traditions and vest flexible mechanisms and market incentives in the guise of formal law.

Unsurprisingly, significant challenges have followed from this transition for administrators and the legislature, and the latter has only succeeded in embracing a more general, preventive stance on environmental protection within the past decade.[48] Attempts to speed up the pace of reform, for instance by supplanting traditional regulation with flexible market instruments, have often been guided by purely theoretical assumptions on the merits of a particular approach, resulting in an overly narrow focus on select mechanisms at the expense of the remaining elements in the policy architecture and the operation of the policy as a whole.[49] As with any process requiring swift adaptation to rapidly changing circumstances, the result has ultimately been characterized by no small amount of tension and outright conflicts.

3.2 Internal and external conflicts – An analytical framework

Generally speaking, one can discern *four* categories of conflicts arising from the introduction of modern climate policies into the existing legal and constitutional order. First, there are *conflicts of objectives*, notably between environmental pro-

---

[44] Martin Winkler, 'Die neue Betreiberpflicht: Klimaschutz und Emissionshandel', 14 *Zeitschrift für Umweltrecht* (2003), pp. 395 et seq., at pp. 395-6.

[45] See Section 1 of the Federal Ambient Pollution Control Act, *supra* n. 25.

[46] Hans D. Jarass, *Bundes-Immissionsschutzgesetz* (6th edn., Munich: C.H. Beck, 2005), Section 1, Marginal annot. 4.

[47] Alexander Reuter and Ralph Busch, 'Einführung eines EU-weiten Emissionshandels – Die Richtlinie 2003/87/EG', 15 *Europäische Zeitschrift für Wirtschaftsrecht* (2004), pp. 39 et seq., at p. 42.

[48] By way of illustration, the precautionary principle has been successively deployed in environmental legislation, as reflected, *inter alia*, in the general duty to prevent harmful environmental effects pursuant to Section 5(1)(2) of the Federal Ambient Pollution Control Act, *supra* n. 25.

[49] Erik Gawel, *Umweltpolitik durch gemischten Instrumenteneinsatz* (Berlin: Duncker & Humblot, 1991), at pp. 2 et seq.

tection and energy market regulation. By way of illustration, the access to electricity grids and minimum feed-in rates guaranteed by the Renewable Energy Sources Act and the Combined Heat and Power Act are conditional on the utilization of specified technologies, with the scope of both acts limited to generation methods defined in the law itself.

In essence, this contradicts the general commitment to free competition set out in energy market legislation pursuant to Community liberalization rules.[50] Likewise, the polluter pays principle adopted as a central tenet of environmental policy as early as 1971[51] is inherently at odds with the current requirement to allocate a significant majority of emission allowances for free to operators under emissions trading rules.[52] Accordingly, the divergent objectives of climate policies and legislation in other issue areas are not always easy to reconcile.

Conflicts can also follow from *divergent regulatory approaches*, notably when conventional rules based on state intervention and 'command and control' meet flexible policies based on the price signals of functioning markets and other financial incentives. An example of such colliding traditions can be seen in the relationship of emissions trading and ambient pollution control, as the former relies on market forces to guide the standard of technology in participating installations, while the latter, in turn, forces rigid performance standards and emission ceilings on each individual operator.[53] By requiring all installations – regardless of cost – to ensure a certain standard of technology, conventional regulation goes against the central premise of emissions trading, given that installations are no longer free to decide whether to acquire further allowances or invest in more efficient facilities.[54] In order to resolve this conflict, the implementation of emissions trading in the European Community necessitated a legislative amendment of pollution control legislation to exempt market participants from the general performance standard.[55]

---

[50] Illustrating the contingency of these promotion schemes, for instance, Section 2(2) of the Renewable Energy Sources Act, *supra* n. 14, formerly excluded hydropower, landfill gas and sewage gas plants with an installed capacity of more than 5 Megawatts and facilities generating electricity from biomass with an installed capacity of more than 20 Megawatts from its scope, a limitation that was lifted in July 2004; see Hermes, 'Instrumentenmix', *supra* n. 39, at p. 581.

[51] The polluter pays principle (*Verursacherprinzip*) was mentioned as one of three guiding principles of German environmental policy in the first environmental action programme of the Federal Government, see Deutscher Bundestag (ed.), *Umweltprogramm der Bundesregierung 1971*, Federal Records of Parliament (BT-Drs.) VI/2719.

[52] See Art. 10 of Directive 2003/87/EC, *supra* n. 26, and, generally, Jonathan R. Nash, 'Too Much Market? Conflict Between Tradable Pollution Allowances and the 'Polluter Pays' Principle', 24 *Harvard Environmental Law Review* (2000), pp. 465 et seq., at pp. 505 et seq.

[53] See, for instance, Section 5(1)(1), (2), and (4) of the Federal Ambient Pollution Control Act, *supra* n. 25, which impose the duty to take protective and preventive action against environmental harm, as well as the duty to use energy efficiently.

[54] Hans-Joachim Koch and Annette Wieneke, 'Das europäische und deutsche Anlagengenehmigungsrecht als Ordnungsrahmen eines Emissionshandels', in Hans-Werner Rengeling (ed.), *Klimaschutz durch Emissionshandel* (Cologne: Heymanns, 2001), pp. 99 et seq., at p. 115.

[55] This amendment affected Council Directive 96/61/EC of 24 September 1996 concerning Integrated Pollution Prevention and Control (IPPC Directive), OJ L 257 of 10 October 1996, pp. 26 et seq., and, in Germany, Section 5(1) of the Federal Ambient Pollution Control Act, *supra* n. 25.

However, similar tensions can also occur between two mechanisms based on the same regulatory premise, exemplified by the way emissions trading interferes with the environmental performance of the Renewable Energy Sources Act. According to an Advisory Council at the Federal Ministry of Economics and Technology, the two incentives virtually cancel each other out as a means of reducing greenhouse gas emissions, given that the generation of electricity with renewable energy sources automatically increases the supply of unused allowances in the trading market and thereby disrupts the price signal required to influence corporate decisions. Moreover, the reductions achieved through renewable energy promotion could be achieved at lower cost if they were left entirely to operators participating in the market rather than a rigid promotion scheme. As the Advisory Council concludes, therefore, the Renewable Energy Sources Act ultimately serves to subsidize European CO_2 emissions originating outside of the German power generation sector, rendering it an environmentally useless, but economically costly instrument 'that should consequently be abolished'.[56]

A further illustration of conflict between two flexible instruments can be discerned in the overlap of emissions trading and the voluntary declaration on climate protection by private enterprise. Under the voluntary declaration, major sectors of German industry had pledged emissions reductions in exchange for a suspension of further regulatory measures;[57] with the introduction of emissions trading throughout Europe, however, the federal government was bound to impose an aggregate limit on emissions for most parties to the agreement. Evidently, this was not in conformity with the reasoning of the earlier arrangement, although the government had no choice in the face of binding supranational commitments.[58]

A third category of frictions can arise when implementing climate legislation in the context of *constitutional doctrines* and *fundamental rights*. On the broader level of constitutional law, the federal organization of legislative and executive powers may impede effective elaboration and enforcement of climate policies in Germany, where a number of relevant issues fall within the purview of the federal legislator, but enforcement and administrative operationalization, in particular, have traditionally been the prerogative of the federate *Länder*.[59] Already, this has caused difficul-

---

[56] See Wissenschaftlicher Beirat beim Bundesministerium für Wirtschaft und Arbeit, 'Förderung erneuerbarer Energien', *supra* n. 38, at pp. 402 et seq.: '[D]as EEG dient der Subventionierung von CO_2-emissionen in Europa außerhalb des deutschen Kraftwerksektors. Der Netto-Effekt des EEG auf die europäischen CO_2-Emissionen ist Null. ... Es wird dann zu einem ökologisch nutzlosen, aber volkswirtschaftliche teurem Instrument und müsste konsequenterweise abgeschafft werden.'

[57] See the Agreement on Climate Protection, *supra* n. 28.

[58] Admittedly, the agreement itself stated that the 'transposition of EU law remains unaffected', allowing for regulation prompted by supranational commitments, see section II on Climate Protection, *supra* n. 28 (translation by the author).

[59] See Section 20 of the Greenhouse Gas Emissions Trading Act, *supra* n. 27, which divides responsibilities between the German Emissions Trading Authority (*Deutsche Emissionshandelsstelle*, or DEHSt) at the Federal Environment Agency and the Ambient Pollution Control authorities of the *Länder*.

ties in the implementation of emissions trading, where administrative responsibilities are held both by a federal agency and local permitting boards, edging dangerously close to a violation of the constitutional preclusion of 'double competences'[60] and resulting in court proceedings to resolve divergent understandings of the law.[61] Also, responding dynamically to changing environmental circumstances may often necessitate the delegation of legislative powers to executive bodies, whereas the German constitution, as shown earlier, requires that important issues attain the democratic legitimacy of statutory law.[62]

Given the universal nature of global warming and the ample scope of mitigating policies, moreover, subjects may be affected in their individual rights and freedoms in manifold ways. Here, again, emissions trading has proven to raise some of the most acute challenges, prompting judicial proceedings and sharp criticism, including accusations that it would incur the 'end of all freedom', entail 'economic coercion by a totalitarian environmental state', and ultimately invite a 'return to the planning economy'.[63] Of particular concern was the ability of emissions trading to affect the constitutionally vested freedom of property and occupation, which also covers economic activity.[64] Opponents of the trading scheme alleged that it would impose an undue burden on market participants already subject to other climate policies and measures, such as energy taxes or performance standards.

Additionally, it was seen to be discriminatory towards sectors covered by the trading scheme, as opposed to other sectors which faced no aggregate emission limits. And altogether, with greenhouse gases traditionally subject to no form of management, the new trading system was held to violate the established balance between individual rights and public concerns, a balance which had found its reflection in the general freedom to engage in pollutant operations subject only to a bound decision of preventive control. Emissions trading, it was argued, would curtail the legal position of operators and render their ability to exercise fundamental rights dependent on a discretionary permit.[65]

Ultimately, these tensions between the objectives and regulatory tradition of pollution control law and emissions trading bred several contentious proceedings,

---

[60] Federal Constitutional Court (*Bundesverfassungsgericht*), judgment of 19 February 2002 (Case 2 BvG 2/00), in 104 Records of the Federal Constitutional Court (BVerfGE), pp. 249 et seq, at p. 266.

[61] See Administrative Court Augsburg, Decision of 1 September 2004, Case Au 4 E 04.1237, 24 *Neue Zeitschrift für Verwaltungsrecht* (2004), p. 1389; Administrative Court Karlsruhe, Decision of 18 October 2004, Case 10 K 2205/04, 25 *Umwelt- und Planungsrecht* (2005), pp. 36 et seq.

[62] See *supra* n. 41.

[63] See Clemens Weidemann, 'Emissionserlaubnis zwischen Markt und Plan – Rechtsstaatsrelevante Probleme des Emissionshandels', 12 *Deutsches Verwaltungsblatt* (2004), pp. 727 et seq., at pp. 728 et seq.

[64] See Arts. 14 and 12 of the Basic Law, *supra* n. 41, respectively.

[65] For an overview of the arguments and their proponents, see my discussion in 'European Emissions Trading and Environmental Regulation in the Member States: Irreconcilable Conflict?', in Teresa Fajardo del Castillo, Christoph Holtwisch, and Tereza Tichá (eds.), *Strengthening European Environmental Law in an Enlarged Union* (Aachen: Shaker, 2004), pp. 162 et seq.

including a suit before the highest administrative court in Germany.[66] And while the latter considered the threshold for a violation of fundamental rights still intact, its determination was based on the free allocation of emission allowances, an outcome that would likely have been different in the event of auctioning and almost certainly would have entailed outraged claims of expropriation. Only auctioning would, however, prevent one of the main shortcomings of emissions trading in Germany, the accrual of substantial 'windfall profits' to electrical utilities, which uniformly passed on the opportunity costs of unsold allowances to consumers.[67]

And finally, tensions may arise between different *regulatory planes*, that is, divergent climate policies in domestic, supranational and international law. What is legal on the domestic plane, for instance, may conflict with precepts of European Community or international law. A salient illustration are all forms of incentives for the promotion of renewable energy sources and energy efficiency measures, as well as the free allocation of allowances to participants in the emissions trading scheme. Depending on the circumstances of the case, such benefits may be classified as state aid under the competition rules of the European Community[68] or as a subsidy under the Agreement on Subsidies and Countervailing Measures (SCM) administered by the World Trade Organization.[69]

While there have been numerous efforts to reconcile separate normative environments by way of conflict or exception clauses, the tedious example of environmentally motivated trade restrictions has shown that institutions tend to prioritize their own agenda at the expense of any competing rules and objectives.[70] A second example is the admissibility of taxes or other charges on bunker fuels for aviation, which – although permissible under domestic law[71] – are precluded by anachronistic exemptions under the Chicago Convention on International Civil Aviation[72] as

---

[66] See Federal Administrative Court *(Bundesverwaltungsgericht)*, judgment of 30 June 2005 (Case 7 C 26.04).

[67] See Helen Lückge and Camilla Bausch, 'Windfall Profits & Wettbewerb in der 2. Handelsperiode', *DowJones TradeNews Emissions* (2006:24), pp. 9 et seq.

[68] See Arts. 87 and 88 of the Treaty Establishing the Economic Community (EC Treaty), as amended by the Treaty of Nice Amending the Treaty on European Union, the Treaties Establishing the European Communities and Certain Related Acts, Nice, 26 January 2001, in force on 1 February 2003, OJ (2001) C 80 of 10 March 2001, pp. 56 et seq.; with regard to emissions trading, see Art. 11(3) of Directive 2003/87/EC, *supra* n. 26, which clearly states that allocation decisions 'shall be in conformity with the requirements of the Treaty, in particular Articles 87 and 88 thereof.'

[69] Agreement on Subsidies and Countervailing Measures, opened for signature 15 April 1994, 1869 *United Nations Treaty Series* (1994) 14.

[70] For an overview, see Sabrina Shaw and Risa Schwartz, 'Trade and Environment in the WTO: State of Play', 36 *Journal of World Trade* (2002), pp. 129 et seq.

[71] Eckhard Pache and Joachim Bielitz, 'Rechtliche Rahmenbedingungen einer Kerosinbesteuerung auf innerstaatlichen Flügen', 16 *Zeitschrift für Umweltrecht* (2004), pp. 297-301.

[72] See Art. 24 of the Convention on International Civil Aviation (Chicago Convention), Montreal, 7 December 1944, in force on 4 April 1947, 15 *United Nations Treaty Series* (1944), pp. 295 et seq., elaborated by International Civil Aviation Organisation, Council Resolution on Environmental Charges and Taxes, adopted by the Council on 9 December 1996 at the 16th meeting of its 149th session, paras. 2 and 4.

well as a number of bilateral agreements, formally known as 'Bilateral Air Service Agreements' (BASAs).[73]

At the European level, moreover, Directive 2003/96/EC calls on member states to

'exempt ... from taxation under conditions which they shall lay down for the purpose of ensuring the correct and straightforward application of such exemptions and of preventing any evasion, avoidance or abuse ... energy products supplied for use as fuel for the purpose of air navigation.'[74]

All this has prevented legislators in Germany and other European states from implementing effective measures to contain emissions from the most rapidly growing source of greenhouse gases,[75] delaying any progress and forcing decision makers to resort to emissions trading as the only permissible measure.[76]

4.   COHERENCE BY DESIGN: ENVISIONING A MANAGEMENT REGIME

As the foregoing section has contended, the legislation implementing German climate policy is currently encumbered by a number of tensions and outright conflicts. While this strategy has by no means proven unsuccessful in reducing greenhouse gases, it could be safely argued that its full mitigation potential has nonetheless been curtailed by such frictions. Against the background of a general reform agenda for the wider body of environmental law, therefore, increased harmonization and simplification within an integrated policy framework also suggest themselves as a possible channel of improved energy and climate regulation, including better delivery of central objectives and principles to often wary addressees.

Of course, a solution at the international or regional level would be preferable for various reasons, notably to lessen the concern about impacts on competitiveness and environmental efficacy. On the international plane, however, the consensus required for a sufficiently ambitious climate regime is currently absent, with the international community already facing challenges in the adoption of fairly moderate targets. At the regional level, in turn, legislative bodies tend to lack the neces-

---

[73] Members of the International Civil Aviation Organisation are required to deposit all such bilateral agreements with the Secretariat, which has compiled the roughly 3000 BASAs in existence in a two-volume collection, ICAO, Document 9511, *Digest of Bilateral Air Transport Agreements* and *Supplement 1*.

[74] See Art. 14(1) of Council Directive 2003/96/EC of 27 October 2003 restructuring the Community Framework for the Taxation of Energy Products and Electricity, OJ (2003) L 283, pp. 51 et seq.; Art. 14(2) of the Directive, however, allows member states to limit the scope of this exemption 'to international and intra-Community transport'. Purely domestic flights, in other words, may be included in a kerosene taxation scheme.

[75] Intergovernmental Panel on Climate Change (IPCC), *Special Report on Aviation and the Global Atmosphere* (Cambridge: IPCC/WMO/UNEP, 1999), especially chapter 6.

[76] See, notably, Proposal for a Directive of the European Parliament and of the Council amending Directive 2003/87/EC so as to Include Aviation Activities in the Scheme for Greenhouse Gas Emission Allowance Trading within the Community of 20 December 2006, COM(2006) 818 final.

sary powers for a comprehensive regulation of greenhouse gases, as is illustrated by the European Union, where political opinion might be more favorable than in an international setting, but the establishing treaty confers no comprehensive power to legislate climate and energy policy.[77] With that in mind, the following sections will outline some considerations relating to the establishment of a domestic scheme to manage greenhouse gas emissions in Germany, starting with the possible sources of a legal mandate, the most important objectives, and tentative design elements.

4.1 The legal context – Identifying a mandate

Before addressing the material objectives and design options of a comprehensive management regime for greenhouse gases, the current legal framework should first be assessed with a view to potential bases for such sweeping reform. In an area as sensitive as energy and climate change, after all, far-reaching policies are likely to find many linkages with fundamental tenets of constitutional law and economic regulation, all of which could impede the adoption of a uniform regime. At the same time, however, the legal order has gradually evolved to accommodate new and increasingly urgent environmental concerns, providing various gateways for a genuine mandate to support the adoption of a stringent climate policy architecture.

First and foremost, the ambitious mitigation objectives agreed to by the federal government provide a strong foundation for comprehensive measures to meet these binding commitments, something a harmonized and consistent strategy is likely to facilitate.[78] Politically, moreover, the agreement forming the parliamentary caucus currently in power affirms a need to continue pursuing a leadership role in climate policy while streamlining the growing body of rules devoted to environmental protection.[79]

And finally, at the level of legal doctrine, one can point to the state objectives of environmental protection and intergenerational sustainability enshrined in the Basic Law,[80] as well as the principle of coherence affirmed by the Federal Constitu-

---

[77] Recently, however, the European Commission has adopted a communication outlining an integrated energy policy, see Communication from the Commission to the European Council and the European Parliament, *An Energy Policy for Europe*, 10 January 2007, COM (2007) 1; likewise, the draft Treaty establishing a Constitution for Europe – rejected after the referenda in France and the Netherlands – would have introduced an independent chapter on energy, giving priority to the promotion of energy efficiency and saving and the development of new and renewable forms of energy, as well as setting out a shared competence for energy policy, see Art. III-256 of the draft Treaty, OJ (2004) C 310, pp. 1 et seq.

[78] See *supra*, section 2.1.

[79] See Sections 7.1-7.3 of the Coalition Agreement of 11 November 2005 between the Christian Democratic Union (CDU), the Christian Social Union (CSU), and the Social Democratic Party (SPD), pp. 65 et seq., published on the Internet at: <www.bundestag.de/aktuell/archiv/2005/koalition/vertrag.pdf> (last accessed on 15 April 2007).

[80] See Art. 20a of the German Basic Law, *supra* n. 41, as amended on 27 October 1994, BGBl. Part I (1994), pp. 3146 et seq., which reads: '[t]he state, aware of its responsibility for present and future generations, shall protect the natural sources of life within the framework of the constitutional order through the legislature and, in accordance with the law and the principles of justice, the executive and the judiciary.'

tional Court, effectively ruling out legislation that stipulates irreconcilable obligations for one and the same addressee.[81] Further support for a harmonized and consistent management scheme may be derived from the principle of integration, which has been vested with the status of positive law by the member states of the European Community,[82] and the principle of proportionality, which could potentially impose a limit on cumulative burdens flowing from the overlap of different measures and policies.[83]

Energy and climate legislation has, to date, been based on the federal power to regulate economic activity as well as, more specifically, ambient air pollution.[84] Given the importance of a common policy framework for uniform economic and legal conditions in Germany, the *Länder* have not been able to challenge the federal power to regulate greenhouse gas mitigation and the various activities involved in energy generation and consumption. Still, climate change differs significantly from traditional pollution challenges, with economic activity widely dependent on energy in its different manifestations, invariably resulting in the emission of greenhouse gases.

Accordingly, there has been ample discussion as to whether the comprehensive management of greenhouse gases automatically incurs a violation of the fundamental right to engage in economic activity, manifested in an alleged right to use air as a resource and a medium for the absorption of emitted greenhouse gases. Indeed, in a decision on the responsibility of the state to compensate damage arising from air pollution, the Federal Constitutional Court observed that

> 'as a medium, "air" is not subject to a management system under public law pursuant to which the holders of basic rights would generally be barred from access, and according to which use would depend on allocation by state bodies subject to their discretion.'[85]

Applied to the context of climate change, such an understanding would preclude the comprehensive management of greenhouse gas emissions within an overarching framework, and would, instead, favor legislation in response to situational threats and narrowly defined issue areas. Unsurprisingly, that very approach has also been

---

[81] In its judgment of 7 May 1998 in Case 2 BvR 1991/95, Records of the Federal Constitutional Court (*BVerfGE*) (1998), pp. 106 et seq., at pp. 118 et seq., the Federal Constitutional Court addressed the permissibility of municipal waste and packaging charges, and found that '[t]he rule of law binds all legislative organs of the Federation and the *Länder* to coordinate their legislation in such a way as to prevent norm addressees from being confronted with countervailing rules which render the legal order contradictory' (translation by the author).

[82] See Art. 6 of the EC Treaty, *supra* n. 68, which reads: 'Environmental protection requirements must be integrated into the definition and implementation of the Community policies and activities referred to in Article 3, in particular with a view to promoting sustainable development.'

[83] On this argument, see Michael Kloepfer, *Umweltrecht* (3rd edn., Munich: C.H. Beck, 2004), Chapter 5 Annot. 284.

[84] See Art. 74(1) Nos. 11 and 24 of the German Basic Law, *supra* n. 41.

[85] See Federal Constitutional Court, Decision of 26 May 1998, Case 1 BvR 180/88, 51 *Neue Juristische Wochenschrift* (1998), pp. 3264 et seq., at p. 3266 (translation by the author).

responsible for the current policy architecture, where individual policies and measures have accumulated without overall coordination, resulting in the conflicts identified in the preceding section.

A very different perception has governed access to water and its pollution, where the affected resource was considered sufficiently vital to justify a comprehensive public management regime (*öffentlich-rechtlich Benutzungsordnung*).[86] With a seminal decision dating back more than twenty years, the Federal Constitutional Court declared that,

> 'because, as a rule, the popular demand for water could formerly be met with an ample water supply, there was no need to restrict the use of groundwater by property owners aside from some provisions designed to protect the interests of neighbours. Growing levels of industrialisation and the establishment of central water supply facilities in cities and municipalities have clearly caused this situation to change, however, necessitating further regulation. (...) Water is one of the most important foundations of all human, animal and plant life. Accordingly, an organised management of water resources is essential, both for civil society as well as the overall economy.'[87]

Based on that understanding, the legislator has been able to introduce a quantitative and qualitative management regime, allowing detailed governance of all activities affecting natural water resources. A similar perception of the climate system and its importance for social and economic welfare could, in turn, serve as the starting point for a comprehensive greenhouse gas management scheme, and there are some initial indications that suggest that the prevailing conception of global warming is shifting.

Called upon to decide a challenge against the emissions trading legislation in Germany, for instance, the Federal Administrative Court has clarified that 'air' could never fall within the ambit of private property, and that, instead, the rules on emissions trading merely regulate the use of property 'insofar as is necessary for the general interest'. In other words, the Court concluded that the emissions trading scheme was an appropriate, necessary and proportional means of protecting the global climate, and that it had merely led to the partial reorganization of that specific area of law without infringing on the vested rights, both nationally and under Community law, of market participants.[88] Given the growing currency and media attention afforded to climate change in recent months, this perception is likely to have become more popular, providing the dogmatic basis for a stringent and comprehensive management of greenhouse gas emissions in Germany.

---

[86] See Sections 3 et seq. of the Water Resources Management Act (*Gesetz zur Ordnung des Wasserhaushalts* – WHG), 19. August 2002, BGBl. I (2002), pp. 3245 et seq.

[87] Federal Constitutional Court, Decision of 15 July 1981, Case 1 BvL 77/78, BVerfGE (1981), pp. 300 et seq., at p. 341.

[88] See, notably, the judgment by the Federal Administrative Court *(Bundesverwaltungsgericht)*, 30 June 2005 (BVerwG 7 C 26.04), affirming that the introduction of emissions trading violated neither European fundamental rights nor the provisions of the Basic Law.

4.2   Integrated Greenhouse Gas management – Clinching the objective

Any attempt to create an overarching framework for the management of green-house gas emissions will subsequently require the definition of uniform policy objectives. Not only is the specification of a common purpose a prerequisite for the determination of substantive principles and regulatory instruments, but its very existence may also have a unifying effect on the subsequent implementation process. Clear objectives have therefore proven essential for effective governance of environmental challenges in the past.[89] Materially, however, these objectives will vary with the substantive scope afforded to the management scheme.

When deciding on the scale of the policy architecture, legislators will be called upon to make a strategic decision on its perimeters. Generally speaking, they can choose to either focus on greenhouse gas emissions and their limitation, or also include broader aspects of energy market regulation and its concurrent aims of energy security and an affordable, competitive energy supply. Although inherently different from mitigation policies, in turn, measures to adapt to global warming could also be included within the ambit of a management regime.

In all cases, however, substantive guidance will follow from the quantitative reduction commitments entered into under international and European Community law,[90] helping define the level of ambition that needs to be pursued with the overall management scheme. By necessity, moreover, a management scheme will have to address central aspects of the energy sector, given that the achievement of the foregoing reduction targets will be conditional on a gradual transition to sustainability through improved efficiency in the exploitation of energy resources as well as in the generation, conversion, distribution, and end-use of energy, but also a shift in the structure of energy sources towards increased use of renewable energy.[91]

Going forward, individual objectives should be specified for each of the foregoing aspects, ideally with quantified targets and timelines. In its recent communication on a Europe energy policy, for instance, the Commission framed an overall 'strategic objective', deferring further operational details to an action plan with quantified targets.[92] In the end, however, these substantive considerations do not prejudice a certain design or scope, leaving decision makers ample freedom in determining what may be politically acceptable and administratively feasible.

Still, if the elaboration of a comprehensive management scheme is also meant to reduce tensions and conflicts between this scheme and other policies as well as

---

[89]   Rudolf Steinberg, *Der ökologische Verfassunsstaat* (Frankfurt am Main: Suhrkamp, 1998), p. 171.

[90]   See *supra*, nn. 10 and 11.

[91]   For an overview, see Martin Jänicke and Tobias Wiesenthal, 'Eckpunkte und Entwicklungslinien einer nachhaltigen Energiewirtschaft', 15 *Zeitschrift für Umweltrecht* (2004), pp. 385 et seq., at p. 385.

[92]   See *supra*, n. 77, with a strategic objective to reduce greenhouse gas emissions by specified amounts, operationalized by quantified targets for energy efficiency improvements, renewable energy promotion, and other aspects of energy demand and supply.

within the scheme itself, it should aspire towards some general objectives of a systemic nature. Altogether, the management scheme should strive for the largest possible degree of consolidation and integration, ensuring the compatibility, consistency and complementarity of its various constituent policies and measures. With normative unity a central condition for the success of greenhouse gas mitigation, individual elements of this strategy must be deployed in conformity with the existing regulatory framework.[93]

By way of illustration, emissions reduction policies should be aligned with energy market rules to avoid tensions between the pursuit of a more sustainable energy supply and further market liberalization. Ultimately, a comprehensive management scheme should avoid sending the contradictory signals relayed by current policies in place in Germany, and instead foster a high degree of harmony in its terminology, substantive goals and principles, and regulatory instruments. Another priority should be placed on curbing the excess regulation of earlier decades, reducing normative complexity and redundant bureaucratic obligations.[94]

Clear, simple and transparent norms may help reduce administrative costs and also promote identification by their addressees, thereby improving the prospects for adequate implementation. Accordingly, a comprehensive management scheme could seek to streamline mandatory procedures and consolidate permitting requirements. Given the dynamic nature of climate change and evolving responses at the regional and international plane, finally, the management scheme should be sufficiently flexible to accommodate external change. In order to safeguard the coherence of the overall scheme, however, future amendments should be subjected to an appropriate assessment procedure designed to identify potential impacts, as should any legislation adopted by administrative entities based on powers conferred on them.[95]

4.3   Towards a Greenhouse Gas Management Act – Specifying Design Elements

In practice, the creation of a comprehensive management regime for greenhouse gas emissions would bring together existing and new instruments under an 'umbrella law', which shall be tentatively designated as the 'Greenhouse Gas Manage-

---

[93] See, for instance, Annex VIII No. 9 of Directive 2005/32/EC of the European Parliament and of the Council of 6 July 2005 establishing a Framework for the Setting of Ecodesign Requirements for Energy-using Products and amending Council Directive 92/42/EEC and Directives 96/57/EC and 2000/55/EC of the European Parliament and of the Council, OJ (2005) L 191, pp. 29 et seq.

[94] Michael Rodi, 'Instrumentenvielfalt und Instrumentenverbund im Umweltrecht', 15 *Zeitschrift für Gesetzgebung* (2000), pp. 231 et seq., at p. 234.

[95] As a suitable model for such an assessment, one might refer to the legislative impact assessment required by Section 44 of the Common Rules of Procedure of the Federal Ministries (*Gemeinsame Geschäftsordnung der Bundesministerien* – GGO), 26 June 2000, Legislative and Ministerial Gazette (*GMBl*) (2000), pp. 525 et seq., as well as the creation of a special institution with the National Norm Review Committee Act (*Gesetz zur Einsetzung eines Nationalen Normenkontrollrates* – NKRG), 14. August 2006, BGBl. I (2006), pp. 1866 et seq.

ment Act' for the purposes of this chapter. As has – to some extent – already occurred with the National Allocation Act (*Zuteilungsgesetz 2007*),[96] albeit without coordination between different legal instruments, such legislation would define overarching management objectives, general principles, common definitions, and mutually coordinated and suitable instruments for the achievement of the specified objectives.

Based on earlier experience with environmental codification in Germany, the Greenhouse Gas Management Act would ideally consist of a general part outlining the shared objectives, definitions, and principles, and a specific part focusing on individual sectors or issue areas, and the measures adopted within its ambit. In the general part, accordingly, the legislator could draw attention to mitigation commitments entered into under international law and specify a global greenhouse gas reduction target, breaking this aggregate objective down to different sectors and activities. General principles could include a duty to take protective and preventive action against climate change, or the duty to use energy efficiently.

As for the selection of suitable instruments, the overall aim should be to arrive at a combination of different instruments capable of influencing individual and collective allocation decisions in line with the objectives defined earlier, and addressing all sources of greenhouse gas emission within the substantive and geographic scope of the Greenhouse Gas Management Act.[97] All instruments currently in use or otherwise discussed for global warming mitigation are theoretically available, including:

- regulations and standards specifying mandatory abatement technologies or minimum requirements for pollution output;
- taxes and charges imposed on undesirable activity by a source;
- tradable permit schemes establishing a limit on aggregate emissions by specified sources and allowing trade among them;
- voluntary agreements between a government authority and one or more private parties with the aim of achieving emissions reductions beyond compliance with regulated obligations;
- subsidies and incentives awarded to an entity for performing a specified action;
- information instruments requiring public disclosure of environmentally-related information, including labeling programmes and rating and certification systems; as well as
- research and development measures involving direct government funding and investment for innovative approaches to mitigation or the infrastructure of emissions reductions.[98]

---

[96] Gesetz über den nationalen Zuteilungsplan für Treibhausgas-Emissionsberechtigungen in der Zuteilungsperiode 2005 bis 2007 (*Zuteilungsgesetz 2007* – ZuG 2007), 26 August 2004, BGBl. I (2004), pp. 2211 et seq.

[97] For a detailed definition of an instrument mix in environmental policy, see Gawel, *supra* n. 49, at pp. 3 et seq.

[98] This list is based on the draft Working Group III contribution to the Intergovernmental Panel on Climate Change Fourth Assessment Report, *Climate Change 2007: Mitigation of Climate Change*

Further instruments might include planning and impact assessment procedures as well as liability rules and criminal sanctions, to name but a few. In order to achieve the strategic objectives of greater consolidation and integration, however, it is imperative that these instruments be carefully screened on the basis of appropriate criteria prior to their inclusion in a Greenhouse Gas Management Act, in order to avoid inconsistencies, conflicts and regulatory overlap.[99] And this is the most challenging stage in the elaboration of a suitable instrument mix. Commonly cited criteria of policy choice, such as those outlined by the Intergovernmental Panel on Climate Change in consecutive Assessment Reports,[100] are generally too formulaic and abstract to allow for the contextuality of selection processes and the manner in which policy instruments are both formulated and implemented within a sophisticated matrix of interests, procedures and institutional mandates as well as material legal constraints.

Accordingly, criteria such as environmental effectiveness, cost effectiveness, distributional considerations and institutional feasibility may provide initial guidance, but are unable to determine the outcome of any given selection process.[101] Additional criteria, such as market conformity, administrative and transaction costs, political acceptance and legitimacy, openness to innovation, and the degree of flexibility and reflexiveness, may also prove helpful, but are equally unable to place the choice of instruments on a purely rational, objective and universally acceptable basis. In that sense, scholars and decision makers will arguably face their most important task when it comes to identifying suitable selection criteria based on the actual necessities at hand, engaging in an interdisciplinary and practically relevant discourse.[102]

5. CONCLUSION AND OUTLOOK

German climate policy has only begun to embrace alternative models of environmental legislation, and it has faced considerable difficulties in doing so. By and large, however, the introduction of economic incentives and market mechanisms has met with regulatory success. A central challenge for the further evolution of

---

(forthcoming, Cambridge University Press), Chapter 13.1.1, available on the Internet at: <http://www.mnp.nl/ipcc/pages_media/FAR4docs/chapters/CH13_Policies.pdf> (last accessed on 1 June 2007), and is by no means comprehensive.

[99] See, for instance, European Commission, *Green Paper on Market-Based Instruments for Environment and Related Policy Purposes*, 28 March 2007, COM(2007) 140, at pp. 8 et seq.

[100] See, most recently, IPCC, *supra* n. 98, at Chapter 13.1.2.

[101] Gawel, *supra* n. 49, at p. 9, affirms that such theoretical criteria suffer from insufficient information on complex chains of causality, physical damage functions, persuasive valuation criteria based on a contingent perception of utility, and macroeconomic costs of reallocating production factors to environmental protection, all rendering such welfare-based approaches to the description of instruments 'at best a general reference system depicting ideal conditions in society' (translation by the author).

[102] See, for instance, Rodi, *supra* n. 94, at p. 241.

climate policy in Germany will lie in ensuring the harmonized operation of a broad basket of different policy instruments. Adoption of a comprehensive Greenhouse Gas Management Act has the potential to streamline the current policy framework and reduce inherent tensions and normative conflicts. Its benefits may also, moreover, reach beyond improved coherence and efficiency. Under a management regime, the ability to emit greenhouse gases would constitute an individually attributable advantage, with fundamental consequences for the scope and intensity of regulation tolerated by constitutional law.

By way of illustration, one might mention two immediate consequences of such a managerial approach: for one, auctioning of emission allowances would face fewer constitutional obstacles, as it would no longer involve placing an unrequited financial burden on operators for activities that were previously unregulated; and next, the introduction of reciprocal emission charges would become permissible under the fiscal rules of the German Basic Law, given that such a charge could now offset an individual benefit of its addressee, the emitter.[103] A comprehensive Greenhouse Gas Management Act would also create a useful framework for the targeted adoption of additional measures in currently unregulated areas (*lacunae*).

Of course, a managerial approach to climate policy may initially run contrary to the established framework for greenhouse gas mitigation, requiring a serious effort from Germany to overcome its rigid and deeply entrenched traditions of environmental regulation. A lengthy convergence process and further legislative amendments will likely become necessary, inviting the possibility of institutional differences and wrangling over legislative powers. Much will ultimately depend on the support of influential interest groups which, in the past, have complained about the *status quo* and then nevertheless opposed any major reform initiatives. Still, a harmonized management scheme offers the largest potential for compliance with ambitious emission reduction objectives, and opens a gateway for urgently needed change along the lines of greater integration and consolidation.

---

[103] For a detailed account, see Nils Meyer-Ohlendorf, Michael Mehling, Astrid Epiney and Stefan Klinski, *Legal Aspects of User Charges on Global Environmental Goods* (Dessau: Umweltbundesamt, 2006), pp. 115 et seq.

THE RUSSIAN FEDERATION AND THE KYOTO PROTOCOL

Wybe Th. Douma* and Daria Ratsiborinskaya**

1. INTRODUCTION

The Russian Federation played a key role in enabling the Kyoto Protocol (KP) to enter into force in 2005. Although the country could benefit a great deal from the Protocol itself, it only ratified after having secured enough additional political incentives from abroad, notably from the European Union. This strenuous process will first be touched upon. Next, the quantitative obligations arising from the KP for Russia will be discussed, and the potential advantages for Russia of some of the Kyoto mechanisms. After that, the general modalities of actually implementating the KP in Russia and the specific legislation with regard to Joint Implementing (JI) adopted at the end of May 2007 will be turned to. Whether the latter legislation will enable a timely start for JI projects in Russia will be the focus point of the last but one part of this chapter. In the final part, some concluding remarks are presented, notably on the prospects of Russia committing itself to a meaningful post-2012 regime.

2. FROM SIGNING TO RATIFICATION

On 16 February 2007 it was two years since the coming into force of the Kyoto Protocol to the UN Framework Convention on Climate Change (UNFCCC). Adopted on 11 December 1997, the Protocol was the first international agreement with quantitative obligations for some of its parties to limit and reduce emissions of greenhouse gases by amounts averaging 5.2% by 2008-2012. By June 2007, 173 countries are now party to the Protocol, including one regional organization, the European Community.

The fact that the Kyoto Protocol could enter into force in the end depended on the Russian Federation. The reason for this was that Article 25 of the Protocol made its entry into force dependent on ratification by not less than 55 Parties to the UNFCCC, incorporating Annex I Parties accounting for in total at least 55% of the total carbon dioxide emissions for 1990 of these Annex I Parties. As the USA (re-

---

* Ph.D., Head of the Department for European Law at the T.M.C. Asser Institute.
** LL.M., Associate researcher T.M.C. Asser Institute and MGIMO.

W.Th. Douma et al., eds., The Kyoto Protocol and Beyond
© 2007, T·M·C·ASSER PRESS, *The Hague, The Netherlands and the Authors*

sponsible for 36.1% of emissions by Annex I countries) decided not to ratify, crossing the threshold depended on Russia, which represented 17.4% of the Annex I countries' emissions.[1] At first, this did not seem to be a problem. In 2001, Russia and the EC anounced that they would work together with a view to the early ratification and entry into force of the Kyoto Protocol.[2] The EC approved the Protocol on 31 May 2002, but Russia needed more time. At the Johannesburg World Summit on Sustainable Development in September 2002, Russia's then Prime Minister Kashyanov did state that 'ratification will occur in the very near future'. At the same venue, Deputy Minister Mukhamed Tsikanov of the Economic Development and Trade Ministry indicated that there was a risk that Russia would not ratify because 'we don't have the economic stimulus, the economic interest in the Kyoto Protocol.' He did add that, for the moment, the plan in Moscow was still to ratify. By March 2003, this had still not occurred, prompting a visit by EC Environment Commissioner Margot Wallström and the Greek and Italian Environment Ministers to encourage Russia to follow through on its pledges. 'The world is waiting for Russia to demonstrate that it is ready and willing to become a major player in the multilateral efforts to combat climate change', Ms Wallström stated ahead of her visit to Moscow.[3]

Between 29 September and 3 October 2003, the World Conference on Climate Change took place in Moscow. It was intended to be the first meeting of the parties to the Kyoto Protocol, but since Russia still had not ratified, this was not possible since the Protocol was not yet in force. President Putin's opening speech made it clear that no exact date had been set for Russian ratification.

> 'Russia is being actively called on to ratify the Kyoto protocol as soon as possible. I am certain that these appeals will also be heard many times at your meeting. I want to say that the Government of the Russian Federation is carefully examining and studying this issue, studying the entire range of complex problems connected with this',

Putin said.

> 'A decision will be made after this work is finished. And, of course, it will be made in accordance with the national interests of the Russian Federation.'

Joke Waller-Hunter, the executive secretary of the UNFCCC, was frank in expressing her disappointment.

---

[1] Note that in 2003, the USA was the world's largest emitter of CO_2, emitting 22% of the total emissions; the 13 EU Euro-zone countries together accounted for 10%; China 16%; Russia 6% and India 5%. The per capita emissions of CO_2 in the USA were 19.9 metric tons; in China 3.2; in Russia 10.3 and in the EU Euro-zone countries 8.2 (World Bank, 2007 Little Green Data Book, covering emission data for the year 2003). China is expected to surpass the USA and become the world's largest emitter of CO2 either in 2007 or in 2008, according to the International Energy Agency (BBC News 24 April 2007).

[2] Brussels EU-Russia Summit of 3 October 2001, Joint Statement, accessed at <http://www.delrus.cec.eu.int/en/p_238.htm> on 22 March 2007.

[3] Press release IP/03/322, Brussels, 5 March 2003.

'We had hoped that [President Putin] would have been somewhat more specific on the date when he would expect the Russian ratification to take place,'

she said.

'Last year in Johannesburg, the [Russian] prime minister announced Russia would ratify the Kyoto Protocol in the nearest future, and we had hoped the nearest future had come today and that we would have a clear signal.'[4]

The Conference itself definitely examined the issue of climate change and a very wide range of problems linked to it, but consensus on the benefits of Russia's ratification of the Kyoto Protocol was not reached. The opponents of Kyoto advanced fierce arguments. Andrei Illarionov, an economic adviser to President Putin, warned, for instance, that 'the Kyoto Protocol will stymie economic growth; it will doom Russia to poverty, weakness and backwardness', claiming that each percentage point of GDP growth is accompanied by a 2% growth in CO_2 emissions.[5] He also claimed that the EU legislation would stand in the way of buying any emission credits from Russia.[6] Later on, Illarionov even went as far as to compare the Kyoto Protocol regime to Auschwitz.[7]

In all likelihood, the extremely negative stance towards Kyoto was part of a strategy to gain concessions from the EC in negotiations on Russia's WTO accession, as Commissioner Verheugen explained in a hearing in the German Parliament.[8] At the EU-Russia summit on 21 May 2004, President Putin announced that

---

[4] <http://unfccc.int/press/stat2003/stat_290903.pdf>.

[5] <http://www.rusnet.nl/news/2003/10/07/print/commentary_01_5423.shtml>.

[6] M. Ebell, *Illarionov explains Russian position on Kyoto Protocol in Washington*, Competitive Enterprise Institute, 13 February 2004. <http://www.cei.org/gencom/014,03867.cfm>.

[7] 'The Kyoto Protocol is a death pact (...) because its main aim is to strangle economic growth and economic activity in countries that accept the protocol's requirements. At first we wanted to call this agreement a kind of international Gosplan, but then we realized Gosplan was much more humane and so we ought to call the Kyoto Protocol an international gulag. In the gulag, though, you got the same ration daily and it didn't get smaller day by day. In the end we had to call the Kyoto Protocol an international Auschwitz.' Illarionov likens Kyoto to Auschwitz, Moscow Times 15 April 2004. In an interview with Jeremy Paxman of the BBC, Illarionov added on 18 May 2004: 'We have received no single argument in favour of this document except political pressure. No link has been established between carbon dioxide emissions and climate change. No other objective facts have been presented in recent time. The IPCC's reports in 1990 and 1995 show it clearly.' and 'We are close to a consensus that the Kyoto Protocol does huge economic, political, social, and ecological damage to the Russian Federation.'

[8] Deutscher Bundestag, Protokoll der 40. Sitzung des Ausschusses für die Angelegenheiten der Europäischen Union, 28 January 2004. Verheugen explained: 'Tatsächlich gebe es Hinweise, dass Russland eine politische Verbindung zwischen dem Abschluss der WTO-Beitrittsverhandlungen und der Ratifizierung des Kyoto-Protokolls herstellen wolle. Russland sehe darin weniger ein formelles Junktim als eine politische Paketlösung. Es scheine eine Herabstufung der Anforderungen an einen russischen WTO Beitritt durch eine Ratifizierung des Kyoto-Protokolls herbeiführen zu wollen.' ('Indeed there are signs that Russia would like to create a political link between the conclusion of the WTO-accession negotiations and the ratification of the Kyoto Protocol. Russia regards this not so much as a formal, legal link but rather as a political package. It seems to be striving for a lowering of the conditions to a Russian WTO accession by ratification of the Kyoto Protocol.')

'[t]he European Union has made some concessions on some points during the negotiations on the WTO. This will inevitably have an impact on our positive attitude to the Kyoto process. We will speed up Russia's movement towards ratifying the Kyoto Protocol.'

Indeed, President Putin decided in favour of the Protocol in September 2004, along with the Russian cabinet. As anticipated after this, ratification by the State Duma (22 October 2004) and Federation Council (lower and upper house of parliament) did not encounter any obstacles.[9] On 4 November 2004 the Protocol was approved by President Putin and Russia officially notified the United Nations of its ratification on 18 November 2004. In Russia itself, the ratification of the Kyoto Protocol was described as a necessity in exchange for the EC's support for Russia's accession to the WTO.[10] Ninety days after Russia's ratification the Kyoto Protocol entered into force on 16 February 2005.

3.   RUSSIA'S OBLIGATIONS AND PROSPECTS UNDER THE KP

The 1997 Kyoto Protocol takes 1990 levels of emissions as its basis. On average, the developed (Annex I) countries to the KP have agreed to ensure a cut in greenhouse gas emissions of at least 5% from 1990 levels in the commitment period 2008-2012. Individual country targets range between -8% and +10%. Russia has been allowed to return its emissions to its 1990 levels. Since 1990, the economies of most countries in the former Soviet Union have collapsed, as have their greenhouse gas emissions. This also holds true for Russia. In 1999, according to Russia's own 2002 inventory report to the UNFCCC, its emissions were 38% below 1990 levels. In its 2006 inventory report, Russia indicated that although its CO_2 emissions grew by 11% between 1999 and 2004, they were still 30% below the 1990 levels. Although Russia's economy is expected to continue growing, the country does not foresee any problems in meeting its quantitative targets under the Kyoto Protocol for the 2008-2012 commitment period.[11] In this respect, it is important to underline how far from the truth Illarionov was when proclaiming that each percentage point of Russia's GDP growth would be accompanied by a 2% growth in CO_2 emissions. In reality, Russia's recovering economy shows that a GDP growth of 30% from 1998 to 2003 was not accompanied by a 60% growth in CO_2 emissions, but rather by a rate well below the GDP growth rate.[12] This partial decoupling

---

[9] The Duma voted by 334 in favour of ratification, with 73 against and two abstentions.

[10] 'Russia forced to ratify Kyoto Protocol to become WTO member', *Pravda*, 2004-10-26. Retrieved on 2006-11-03. Whether one follows the EU or the Russian view on the reasons for ratification, it is definitely misleading to say that 'Moscow offered prompt and unambiguous support to relevant European polices, easily ratifying the Kyoto Protocol' (B. Kaimakov, 'G8 summit promises nothing sensational', *RIA Novosti*, 2007-06-01. Retrieved on 2007-06-03).

[11] Fourth National Communication to UNFCCC and Progress Report submitted to UNFCCC on 13 February 2007.

[12] Institute of Energy Investigation of the Russian Academy of Sciences, Moscow, as cited in A. Korppoo, J. Karas and M. Grubb, *Russia and the Kyoto Protocol, Opportunities and Challenges*, Chatham House, 2006 at p. 25.

of GDP and carbon emissions since 1998 shows that energy efficiency in Russia is increasing, in line with the Russian Energy Strategy.[13]

The latest Russian forecast is that its CO_2 emissions will still be some 25% below the 1990 levels by the end of 2012, in spite of further economic growth.[14] This means that, in all probability, Russia will have no problems in meeting its quantitative obligations under the Kyoto Protocol.

As for the post-2012 period, for which new obligations are still to be agreed upon, Russian forecasts predict that by the year 2020, CO_2 emissions will still be 10 to 20% below 1990 levels.[15]

An interesting point that was made in favour of a vigorous Russian energy-efficiency policy is the potential for energy exports. Russia's economic growth over the last few years has been due, to a considerable extent, to its oil and gas exports. One commentator, Igor Bashmakov of CENEf (Centre for Energy Efficiency), even claims that if Russia would double its GDP (as is the goal set by President Putin) without improvements to its energy intensity levels, it would already lose its capacity to export oil and gas by the year 2010.[16] In other words: if Russia wants to keep up its economic growth, this will depend on energy exports and these are only possible if energy efficiency increases.

Whether or not Russia will be able to benefit from the Kyoto Protocol is doubted by some,[17] but most agree that joining the Protocol offers valuable options for Russia,[18] notably because its industry and electricity production suffer from major energy inefficiencies that could be dealt with via Joint Implementation projects. There is a large potential for improving energy efficiency in Russia,[19] with energy use per unit of GDP being much greater than in the EU.[20]

---

[13] <http://www.rg.ru/2003/10/07/energetika.html>. A Summary of the Energy Strategy of Russia for the Period up to 2020, Ministry of Energy of the Russian Federation, Moscow, March 2003 is available at <http://ec.europa.eu/energy/russia/events/doc/2003_strategy_2020_en.pdf>. The document sets out Russia's energy policy for the period up to 2020. It was approved on 23 May 2003 and confirmed by the government on 28 August 2003. The main goals of this Strategy are energy safety, energy effectiveness, budget effectiveness and ecological energy security.

[14] Progress Report submitted to UNFCCC on 13 February 2007, p. 12.

[15] Idem.

[16] Bashmakov, I. (2004), 'Russian GDP doubling, district heating and climate change mitigation', presentation at the UNFCCC Workshop: Climate Change Mitigation: Vulnerability and Risk, Sustainable Development, Opportunities and Solutions, 19 June 2004, Bonn, Germany. Accessed at <http://www.cenef.ru/bulletin/Bonn.ppt> on 15 May 2007.

[17] Notably by the already mentioned former presidential adviser A. Illarionov.

[18] See for instance Kommersant 22 January 2007 and A.S. Dagoumas, G.K. Papagiannis, P.S. Dokopoulos,'An economic assessment of the Kyoto Protocol application', Energy Policy 34 (2006) 26–39.

[19] See for instance <http://ec.europa.eu/energy/russia/reference_texts/doc/2006_10_energy_efficiency_en.pdf>.

[20] Viktor Danilov-Danilyan, 'Russia, Kyoto Protocol and Climate Change', Novosti, 16 February 2007. In 2003, Russia's CO_2 emissions per unit of GDP were 1.2, much higher than in the Eurozone countries with 0.3; the USA 0.6 and China 0.6. The units reflect CO_2 emissions in kilograms per unit of 2000 GDP in purchasing power parity (PPP) terms, i.e., gross domestic product converted to international dollars using PPP rates. An international dollar has the same purchasing power over GDP as a US dollar has in the United States. Source: Worldbank, Little Green Data Book 2007.

Besides JI, to which we will return below, another option under the KP is the selling of assigned amounts units (AAUs). The KP enables countries like Russia with a surplus to exchange parts of this surplus to other Annex I Parties. These unused assigned amounts are often referred to as 'hot air' because purchases would not yield environmental benefits. For Russia, however, the surpluses are a result of the humiliating economic decline and the restructuring process, and benefits from sales of these surpluses are probably regarded more as compensation than as a 'windfall'.[21] Anyway, Russia's significant surplus of emission quotas can be sold to other countries with targets when international emissions trading starts. The amount of money that Russia could earn by selling emission credits under the Kyoto Protocol will depend on the total amount of quotas it decides to sell and the market price of such quotas. Without the USA participating, demand and thus prices are probably going to be less than first anticipated. Potential buyers of emission credits from Russia are Canada, Japan and the EU. Russia itself introduced the concept of a Green Investment Scheme (GIS) at COP-6 in December 2000. With such a GIS in place, revenues from the sale of surplus allowances would be earmarked for environmentally-related purposes. The idea was not elaborated any further pending the period in which Russia was dragging its feet in ratifying the KP, but could be revived to accommodate concerns about 'hot air' trading.[22]

As already indicated above, Russia could also invite other countries with emission targets to carry out emission reductions projects in Russia from which emission credits could be sold. Such Joint Implementation (JI) projects would enable Russian industry, especially in the energy sector, to become more efficient and less pollution-intensive. The projects could also help industry and municipalities to modernize and acquire new technology, for example in the district heating sector.[23] Before JI projects can start in Russia, a number of institutional, as well as reporting and accounting requirements set out in the KP and the Marrakech Accords[24] need to be met.

4. THE ADOPTION OF GENERAL IMPLEMENTATION MEASURES

Countries that have ratified the Kyoto Protocol must create institutional arrangements for implementation. In order to prepare for the first commitment period of the Kyoto Protocol (2008-2012), after its ratification in November 2004, Russia

---

[21] See further A. Korppoo, J. Karas and M. Grubb, 'Russia and the Kyoto Protocol, Opportunities and Challenges', Chatham House, 2006 at p. 41 and E. Woerdman (2005), 'Hot air trading under the Kyoto Protocol: an environmental problem or not?', *European Environmental Law Review*, 14(3), 71–77.

[22] The introduction of a GIS package is currently under development in Ukraine, Bulgaria, Romania, and Latvia.

[23] Q&A: The entry into force of the Kyoto Protocol, MEMO/04/245, Brussels, 22 October 2004.

[24] Decisions adopted by the Conference of the Parties serving as the meeting of the Parties to the Kyoto Protocol at its first session (COP/MOP1), FCCC/KP/CMP/2005/8/Add.2.

first of all adopted a National Action Plan on the Kyoto Protocol Implementation in March 2005.[25] The plan sets out the distribution of responsibilities among 15 Federal governmental bodies (Ministries and Agencies) with regard to the realization of the KP in Russia (by mentioning the lead body and those further involved) and a (overly optimistic) timepath by which further action is to be taken (for instance: adopting JI procedures by mid-2005). In May of the same year, an Inter-Agency Commission on the Kyoto Protocol was established (led by Dept. Minister A. Sharonov of MEDT).[26] This Commission upgraded the aforementioned Action Plan in July and October 2005. The MEDT emerged as the responsible ministry coordinating Russia's implementation of the Kyoto Protocol.

Issues that have been dealt with since then include:

- A national system of emissions estimation (национальная система оценки выбросов парниковых газов) under Article 5 Kyoto Protocol; established by Government Decree (01.03.2006 № 278-р).[27] Roshydromet (the Russian HydroMeteorological organisation) is responsible for carrying this out.
- A National Inventory of emissions and removals (национальный кадастр выбросов и абсорбции) under Article 7 Kyoto Protocol; established by Roshydromet Order (30.06.2006 No. 141).[28] On the basis of this document the National Inventory Report on 1990-2004 emissions was complied.[29]
- The Fourth National Communication under the UNFCCC (2006).[30]
- The Report on progress in the implementation of the Kyoto Protocol (prepared by the Ministry of Economic Development and Trade).[31]
- The Report on fixed quantities.[32]

---

[25] Комплексного плана действий по реализации Киотского протокола в Российской Федерации. The National Action Plan was actually initiated by a government decision of 30 September 2004.

[26] Межведомственная комиссия по проблемам реализации Киотского протокола в Российской Федерации (МВК).

[27] Правительство Российской Федерации, Распоряжение от 01.03.2006 N 278-р [О создании российской системы оценки антропогенных выбросов из источников и абсорбции поглотителями парниковых газов, не регулируемых Монреальским протоколом по веществам, разрушающим озоновый слой]

[28] «Об утверждении Порядка формирования и функционирования российской системы оценки антропогенных выбросов из источников и абсорбции поглотителями парниковых газов». Приказ Росгидромета от 30 июня 2006 г. N 141.

[29] On 8 January 2007 Russia presented to the UNFCCC secretariat its National Inventory Report for the period 1990-2004. Compared to the Fourth National Communication data, presented in October 2006, emissions of the base year 1990 according to the new inventarization have increased to 576 million tonnes of CO_2-equivalent. The major reason for this data increase is better access to information and the use of more exact data relating to production levels in the energy sector. The information was presented by the Federal Service of state statistics to Roshydromet, which is responsible for emission inventarisation.

[30] As of 10.03.2007 available in Russian only.

[31] Доклад об очевидном прогрессе по выполнению обязательств по Киотскому протоколу (Минэкономразвития России), <http://www.ncsf.ru/resources/materials/62.pdf>, retrieved on 10 June 2007.

[32] Доклад об установленном количестве (МПР России), not yet available on the internet.

- Organization administering the Russian CO_2 register – 'Federal centre of geoecological systems'; Government decree of 15.12.2006 No. 1741-p[33] on the creation of an organization-administrator of the Russian cadastre of carbon emissions.

In many respects, the reporting by Russia up until the Third National Communication was judged to be insufficient and not in accordance with prescribed formats, although progress was visible over the years. Hopefully, the Fourth National Communication, and the National Inventory Report for that matter, will continue to be more precise than the previous reports.

5. THE JOINT IMPLEMENTATION REGIME

Where JI is concerned, two options are available for the Parties to the KP: Track 1 (full eligibility) and Track 2 (partial eligibility). At the time of writing, Russia does not meet full eligibility and has therefore opted for the second option, which implies, among other things, that JI projects will need to be verified by an Accredited Independent Entity (AIE), accredited by the JI Supervisory Committee (JISC). For quite some time, Russian projects had been seeking approval from the Russian administration. On 28 May 2007, the long awaited Decree on JI rules in Russia was finally adopted (2 years later than the Russian Action Plan had set out).[34] The main points of this decree can be summarized as following:

- The Russian JI Coordinating Centre responsible for the preparation of JI project approval is based at the Ministry of Economic Development and Trade (MEDT);
- The list of projects chosen by the MEDT is sent to the Government where it receives final approval in consultation with the Federal executive bodies (governing given activities);
- The Russian Government can dismiss approved projects for reasons such as exceeding deadlines in reporting to the Coordination Centre and a discontinuation of the activities of a private entrepreneur.

In pursuance of this decree, the *Rules on approval and assessment of JI project realization No. 0796* were adopted on 30 May 2007.[35] These include the following provisions:

---

[33] Распоряжение Правительства РФ от 15 декабря 2006 г. N 1741-p о создании организации-администратора российского реестра углеродных единиц.

[34] Постановление от 28 мая 2007 г. N 332 О порядке утверждения и проверки хода реализации проектов, осуществляемых в соответствии со статьей 6 Киотского протокола к Рамочной конвенции ООН об изменении климата (Decree of 28 May 2007 N 332 on the regime for the approval and control of the operation of projects, carried out in accordance with Article 6 of the Kyoto Protocol to the UNFCCC).

[35] Положение от 30 мая 2007 г.,N 0796 об утверждении и проверке хода реализации проектов, осуществляемых в соответствии со статьей 6 Киотского протокола к Рамочной конвенции ООН об изменении климата.

- The deponent deposits a project concept, project documentation and project passport with the coordinating centre (the Ministry); The documents shall be prepared according to the templates available on the website of the coordinating centre and signed by the deponent;
- The concept is approved by the commission set up by the coordinating centre, which also chooses independent expert organizations using the criteria given in Article 6 of the Kyoto Protocol. The final list of expert organizations shall be available on the website of the coordinating centre;
- Foreign deponents shall deposit the documents in the language of the respective country, with a notarized translation in Russian;
- It is possible to conclude an international agreement on JI project implementation with a foreign counterpart;
- Projects' realization shall not take place before January 1, 2008 or after December 31, 2012;
- It takes 10 working days for the coordination centre to register the project concept deposition, 30 days for the respective Federal body to deal with the concept and to provide a positive or negative conclusion thereon, and 10 working days for the Government to give its final approval to the list of JI projects;
- The JI projects' realization is assessed by a designated body (this differs according to the project's implementation sector), which on certain grounds (an incorrect project implementation) can come up with suggestions to the coordination centre as to the project's termination;
- The reporting period on the project is one calendar year.
- The disputes related to project realization are settled by means of negotiations within 6 months; otherwise the parties can resort to the courts or to the court of arbitration in accordance with the legislation and the international obligations of the Russian Federation.

Not everything is yet ready for JI projects in Russia. The MEDT, together with the Ministry of Foreign Affairs, will by 1 September 2007 need to come up with the template of international agreements between the Government of the Russian Federation and foreign governments on project realization. Futhermore, standards on project efficiency will need to be adopted for project approval. The MEDT could have these ready by the end of the summer of 2007. Also, the list of independent experts needs to be made available. Thus, it remains unclear when exactly JI projects can go ahead in Russia. Also, the decree and rules described above still leave several questions unanswered, notably which other organizations will be involved, and thus a clear and transparent approval system for JI projects is still lacking. Meanwhile, the timeframe for JI projects is becoming really urgent as the start of the first commitment period (1 January 2008) is approaching and the finalization of the rules and guidelines with regard to JI procedures will probably take up the rest of the summer of 2007, at least. At any rate, the die is cast. As is so often the case in Russia, the chances are that, somehow, things will work out.

6. CONCLUDING REMARKS

Russia took its time before ratifying the 1997 Kyoto Protocol. After doing so in November 2004, an intensive debate on the modalities of implementation started. In March 2005, a National Action Plan on the Kyoto Protocol's Implementation was adopted, while in May of the same year an Inter-Agency Commission on the Kyoto Protocol was established. Only at the end of May 2007 was the first legislation adopted with regard to JI. This means that the country's first emission reduction projects could soon be submitted to the UNFCCC JI approval process. The UNFCCC Secretariat estimates that these first emission reduction projects in Russia could reduce emissions of greenhouse gases by more than 65 million tonnes of CO_2-equivalent over the five years from 2008 – i.e., approximately the emissions of Sweden during one year.[36] The Ministry of Economic Development and Trade (MEDT) will be taking the lead in assessing, controlling and implementing the JI projects. On 15 June 2007, Deputy Minister Andrey Sharonov of MEDT expressed his confidence that with the help of the EU as a 'main partner', Joint Implementation will help Russian enterprises to increase energy efficiency 'for relatively little money'. He stated that cutting greenhouse gas emissions in Russia is less expensive than in countries that belong to the Organization for Economic Cooperation and Development because Russia is starting from a lower base, and explained that because of low energy efficiency and aging equipment, cutting a ton of emissions costs from $5 to $25 in Russia, compared with from $50 to $100 in the OECD countries. Sharanov also indicated that some 29 projects for improving the emissions standards at domestic companies have already been submitted to the Kyoto secretariat for approval. The average cost of these projects is from 10 million euros ($13.3 million) to 50 million euros. The power industry accounts for a quarter of Russia's greenhouse gas emissions, according to Unified Energy Systems chief Anatoly Chubais. UES is seeking to raise $1 billion by selling emission credits gained from greenhouse gas reduction projects such as coal-burning surges.[37]

Hopefully, the implementation of the Kyoto Protocol will indeed turn out to be profitable for Russia. That would certainly contribute to a more positive stance of the country to a post-2012 regime than the views that have been expressed so far.

For a start, at a large conference on Climate Change in St. Petersburg held at the end of May 2007, the only state official attending, Federation Council Speaker Sergej Mironov, claimed in his opening speech that global warming does not exist and that, in fact, the world is getting cooler. At the same conference, a government energy adviser warned that the KP would threaten the independence of the Russian oil and gas industry.[38] More importantly, at the G8 Summit of Heiligendamm at the

---

[36] UNFCCC press release 1 June 2007, 'Russia poised to make use of Kyoto Protocol mechanism'.

[37] M. Levitiv, 'Russia seeks EU help on emissions', *Moscow Times* 18 June 2007, p. 6.

[38] S. Shuster, 'Mironov claims world is getting cooler', *The St. Petersburg Times*, 29 May 2007. Reportedly, Mironov based this remarkable opinion on several obscure Russian studies and – much in the style of Illarionov at conferences personally attended by the authors – hurried out at the end of the session without offering those attending the opportunity to ask questions.

start of June 2007, Russia – together with the USA – refused to agree on firm post-2012 commitments. Instead, Russia and the USA only accepted that they would 'consider seriously' joining the other G8 group members in their efforts in combating climate change by cutting greenhouse gas emissions by 50% by 2050.[39] Getting Russia on board a meaningful post-2012 regime will thus be among the many challenges awaiting the European Union and others. Hopefully, the lessons that can be learned from Russia's ratification process will be taken to heart.

---

[39] G8 Summit 2007 (Heiligendamm), Summit Declaration 7 June 2007: 'In setting a global goal for emissions reductions in the process we have agreed today involving all major emitters, we will consider seriously the decisions made by the European Union, Canada and Japan which include at least a halving of global emissions by 2050. We commit to achieving these goals and invite the major emerging economies to join us in this endeavour.' (p. 15/16)

LEGAL AND INSTITUTIONAL BARRIERS TO KYOTO PROTOCOL IMPLEMENTATION IN NON-ANNEX I COUNTRIES IN SOUTH-EASTERN EUROPE AND CIS

Marina Olshanskaya*

ABSTRACT

South-Eastern Europe and CIS countries boast a significant, but still untapped potential for a cost-effective reduction of greenhouse gas (GHG) emissions and participation in the Clean Development Mechanism (CDM). The establishment of a national institutional framework for the Kyoto Protocol, such as Designated National Authorities (DNA), is a first and absolutely necessary step for the countries wishing to host CDM projects and to attract additional financing for GHG mitigation/sequestration projects. In this region the progress in designing national institutional frameworks for the Kyoto Protocol has been uneven. While new EU member states and new accession countries (Bulgaria and Romania) have moved forward with creating their JI Secretariats and mobilizing resources for capacity building, this process has been rather slow in the rest of the region, especially on the side of non-Annex I countries in South-Eastern Europe and the CIS (Central Asia, Southern Caucasus and Moldova). For most of the countries in the CIS and Western Balkans, the absence of a fully defined DNA stems from a mix of factors, including a lack of understanding of the requirements of the CDM, limited financial resources for training and allocating government personnel for project review, and an absence of significant technical assistance from the donor community. In order to identify best practices, lessons learnt and priority capacity building needs, the United Nations Development Programme (UNDP) recently initiated the preparation of a regional study on existing institutional frameworks for the Kyoto Protocol (UNDP BRC 2006). This paper highlights the most important findings of the study and builds on the results of the UNDP-supported regional capacity building programme for the Kyoto Protocol in Eastern Europe and the CIS.

1. GHG REDUCTION POTENTIAL AND THE STATUS OF THE CDM PIPELINE

High GHG intensity is a distinguishing feature of most CIS (Commonwealth of Independent States) and South-Eastern European economies. Six CIS countries are among 20 of the most GHG-intensive economies globally: Uzbekistan, Kazakhstan and Turkmenistan are the most prominent in this regard, ranked as the first, fourth and fifth GHG-intensive countries in the world (see Table 1). The high ratio of

---

* UNDP Regional Center for Europe and CIS.

W.Th. Douma et al., eds., The Kyoto Protocol and Beyond
© 2007, T·M·C·ASSER PRESS, *The Hague, The Netherlands and the Authors*

GHG emissions to economic output in former socialist economies signals the high cost-effectiveness of potential CDM projects as it implies that a large volume of GHG emission reductions can be achieved per 1 US$ of investment.

CIS and South-Eastern European non-Annex I countries, however, have been very slow in capitalizing on their significant GHG reduction potential and their share in the carbon market remains insignificant. Only five of the 421 CDM projects registered by the CDM Executive Board by December 2006 are from the CIS (there are no registered projects from South-Eastern Europe). Altogether non-Annex I countries in CIS and South-Eastern Europe account for 1.1% of the global CDM pipeline and provide for only 0.3% of the CER volume expected to be delivered by 2012, lagging behind even other carbon market outsiders, such as Sub-Saharan Africa (4.9%) or North Africa and the Middle East (2.0%).

Table 1. Carbon Intensity in selected non-Annex I CIS Countries

Country	Carbon intensity of GDP, tCO2eq/mln$intl	World rank
Uzbekistan	3,079.0	1
Kazakhstan	1,787.5	4
Turkmenistan	1,174.8	5
Moldova	1,134.9	12
Azerbaijan	1,112.7	13
Tajikistan	895.1	18

Source: CAIT 2006

1.1   Countries in a 'pre-market' phase: Central Asia and South-Eastern Europe

These countries are characterized by minimum, if any, engagement in carbon market transactions. They lack basic institutional structures, such as the Designated National Authorities (DNAs), and some have not yet ratified the Kyoto Protocol. Nevertheless, many of them boast significant, yet untapped potential for a cost-effective reduction of GHG emission (Serbia in South-Eastern Europe, Turkmenistan and Uzbekistan in Central Asia). Key institutional barriers that these countries face are summarized below.

First, the level of awareness and understanding of carbon finance and the Kyoto Protocol, particularly among high-level decision makers and sectoral agencies, remain extremely low. Responsibility for Kyoto Protocol implementation in most cases has been delegated to the Ministries of Environment due to their mandate in the area of climate change and UNFCCC. As a result, other key governmental and private sector actors might have been left uninformed about the opportunities that the emerging carbon market can provide for implementation of their investment plans and policies.

Secondly, the lack of data, information and analysis poses serious barriers to the identification of sectors, industries and GHG-reduction opportunities for carbon finance. Even in the countries where political support does exist, national actors often lack a basic understanding of how and where to look for opportunities for carbon finance. There is a need, especially in GHG-rich industries and sectors, to carry out an inventory of their investment and modernization plans in the context of carbon finance criteria and eligibility requirements to identify projects that offer good CDM prospects under current market factors (i.e., a limited time-frame for development and implementation, the costs of Certified Emission Reductions (CERs), the availability of technology/investments).

The absence of the required institutional framework and national governance structures to support carbon finance transactions is yet another important obstacle to the dynamic development of carbon markets in these countries. A prerequisite for the developing countries to access CDM is the need to ratify the Kyoto Protocol and establish DNA. This should be complemented by the establishment of a more conducive legal and regulatory environment for carbon finance, including special legal, fiscal and accounting provisions for new types of market commodities (CERs) and deals (ERPAs). This paper will further illustrate that few countries have so far managed to fully meet these requirements.

1.2   Countries in an initial stage of engagement in the carbon market: Southern Caucasus and Moldova

The second category of non-Annex I countries in South-Eastern Europe and the CIS are those that have already gained some initial experience and understanding of the carbon market and have identified and started the development of prospective CDM projects. The latter, however, in most cases fall under the category of 'low-hanging fruits', i.e., they offer high and very cost-effective GHG emission reduction potential, are easy to develop and implement, but often have little, if any, impact on a host country's sustainable development. There are several barriers to a more sustainable development-oriented carbon market which can be summarized as follows.

As illustrated in Table 2 carbon finance alone in many cases cannot generate a sufficient source of revenue to make the underlying projects economically attractive. This is particularly relevant for projects aimed at energy efficiency improvement, the promotion of renewable energy and bio-sequestration, which typically have strong sustainable development benefits, but relatively low (or more risky) carbon benefits. A review of projects in the regional CDM pipeline (UNEP Riso 2007) reveals that only in one case (Optimization of cement production in Armenia) is the impact of CDM registration on project profitability such that it became sufficiently attractive for investors (i.e., IRR – 36%), while for the rest of the reviewed projects the impact of CER revenues on the projects' Internal Rate of Return (IRR) is largely insufficient to bring it to a level which is comparable with financers' hurdle rate. This analysis also shows that the increase in IRR is particularly marginal (1-2%) for projects with a high sustainable development dividend, such as small-scale hydro-power in Armenia, wind in Cyprus or afforestation activities in Moldova (i.e., the last-mentioned will not become financially viable, even taking into account the sales of CERs, in the next 100 years). Such unfavorable economics of CDM projects is considered as a serious market limitation and dictates the need to combine a variety of sources, including official development assistance (ODA), governmental support, carbon or private investment, and to adopt a more creative and complex approach to structuring project financing for CDM.

Table 2. Impact of CER revenues on IRR of CDM projects

Sector	Project Title	IRR/NPV without carbon revenues	IRR/NPV with carbon revenues	Incremental IRR/NPV
Agriculture	Lusakert Biogas Plant (LBP): methane capture and combustion from poultry manure treatment	negative	13%	13%
Landfill gas	Nubarashen Landfill Gas Capture and Power Generation Project in Yerevan, Armenia	n/a	15%	15%
Landfill gas	Landfill Gas Capture and Power Generation Project in Tbilisi, Georgia	negative	6.33%	6%
Hydro power	Argichi Small Hydroelectric CDM project, Armenia	10.5%	13%	2.5%
Cement	Optimization of cement production technological processes at CJSC 'MICA-CEMENT', Armenia	24%	36%	12%
Wind Power	Mari Wind Farm Project, Cyprus	n/a		1.77%
Wind Power	Alexigros Wind Farm Project, Cyprus	n/a		2.02%
Forestry	Moldova Soil Conservation Project	4.2%	5.8%	1.6%

Source: Internal rates of return from selected draft PDDs submitted along with proposed new methodologies to UNFCCC (available at <http://cdm.unfccc.int/Projects>)

The absence of reliable data and approved methodologies has been identified as another obstacle to a wider project representation in the regional CDM pipeline. As reported in UNDP's recent CDM Assessment report (UNDP 2006), nearly half the CDM sector categories have only few methodologies (e.g., energy distribution, transport, forestry) or no approved methodologies at all (e.g., district heating); whereas it is exactly these sectors (energy efficiency and forestry) that possess the highest potential for cost-effective GHG reduction in the South-Eastern and CIS countries. The development of CDM methodologies and finding data for baseline studies for projects in these sectors proved to be extremely challenging, time-consuming and resource-intensive. Without additional methodological support it will remain diffi-cult to leverage carbon finance to such sectors as energy efficiency, transport or forestry.

Finally, as the carbon market evolves, small and medium-sized countries, like the majority of South-Eastern European and the CIS countries, will find themselves in a less advantageous market situation than larger countries, such as India or China. Due to the size and structure of their economies, small and medium-sized CDM hosts have a scarce supply of projects with a high volume of CERs. On the other hand, a number of validated small-scale CDM projects exist which have still failed to find an interested investor, or an available offer does not compensate for the transaction costs. To correct such imbalances, ODA support is required to bundle small-scale activities and their aggregation for collective marketing.

2. STATUS OF THE NATIONAL INSTITUTIONAL FRAMEWORK IN EASTERN EUROPE AND THE CIS

The development of DNAs and the institutional capacity to evaluate and approve CDM projects in Southern Europe and the CIS has been slow compared with other regions of the world. Of the 14 countries surveyed (Armenia, Albania, Azerbaijan, Bosnia and Herzegovina, Georgia, FYR of Macedonia, Moldova, Kazakhstan, Kyrgyzstan, Serbia, Montenegro, Tajikistan, Turkmenistan, and Uzbekistan) eight have designated a DNA contact point (see Table 3). However, the notification of a DNA contact point to the UNFCCC does not necessarily mean that the country in question has developed an operational DNA with the institutional infrastructure required to satisfy major DNA functions, such as the establishment of rules for project eligibility and approval. Of the six countries with a DNA contact point, Moldova has established a fully-fledged DNA with almost complete project criteria and approval procedures in place. Armenia, Georgia, and Uzbekistan are moving quickly towards setting up their DNA. Armenia and Georgia have nominated their Ministries of Environment and Uzbekistan has nominated its Ministry of Economy to act as DNAs and have prepared the draft national CDM approval procedures which are currently awaiting official governmental endorsement. Kazakhstan has developed unified guidelines for project submission and has approved two projects for JI/CDM,[1] but has not yet agreed on a final institutional framework for project review and is still debating possible structures depending on whether the govern-

---

[1] During the creation of the Kyoto Protocol, Kazakhstan negotiated a special agreement, whereby it would participate in the Kyoto Protocol as an Annex I country. This means that once Kazakhstan ratifies the Kyoto Protocol, the country voluntarily takes on a binding GHG emission target and is eligible for participation in Joint Implementation (JI). JI is the CDM-equivalent mecha-nism for jointly developed GHG projects and only these projects are developed jointly between An-nex I countries. Because of the country's special status, projects approved by the Government of Kazakhstan are evaluated as both JI and CDM projects, but follow the general procedures for CDM project review and submission.

ment decides to ratify the Kyoto Protocol as Annex I or non-Annex I. Tajikistan is the furthest behind in the CIS region and is still in the process of ratifying the Kyoto Protocol.

With only three Kyoto Protocol ratifications (Macedonia, Montenegro and Albania), countries in South-Eastern Europe are furthest behind in the CDM institutional development process. Macedonia and Albania with the UNDP's technical assistance started the establishment of their DNA in late 2005; in both countries the Ministries of Environment have been authorized to act as national DNAs. Serbia and Bosnia and Herzegovina (BiH) remain on the fence regarding ratification, even though this will be a requirement for their integration into the EU. A ratification package was prepared by Serbia and BiH and in Serbia was submitted to Parliament. Both packages include recommendations for setting up a DNA.

Table 3. Legal Structure of DNAs in South-Eastern Europe and the CIS

Country	DNA Reported to the UNFCCC	DNA Legal Structure	DNA Administrative Unit/Secretariat
Albania	Ministry of Environment	*In process*	UNDP Climate Change Umbrella Programme, Ministry of Environment
Armenia	Ministry of Nature Protection	**Ministry of Nature Protection**	*Working group to be established.* Currently 2 contact points in Ministry of Nature Protection: 1) UNFCCC Focal Point/Head of Environmental Protection Department, and 2) Head of International Cooperation Department
Azerbaijan	Ministry of Ecology and Natural Resources	*In process*	
Bosnia and Herzegovina	--	--	--
Georgia	Ministry of Environment Protection and Natural Resources	**Ad hoc CDM National Board** of high-level officials from sectoral ministries and headed by the Min. of Environment.	Climate Change Office in Ministry of Environment Protection and Natural Resources
Macedonia	Ministry of Environment and Physical Planning	Ministry of Environment and Physical Planning (MoEPP). MoEPP has final legal authority on project review, approval, and signature of the host country endorsement and/or approval letter. Other relevant ministries (Ministry of Economy,	The DNA Secretariat is housed at the MoEPP, within the Department of Sustainable Development.

Table 3. Cont.

Country	DNA Reported to the UNFCCC	DNA Legal Structure	DNA Administrative Unit/Secretariat
		(Ministry of Economy, Agriculture and Water, Finance, Transport and Communication) will be involved in project review through a two-step review process.	
Moldova, Republic of	The State Hydrometeorological Service of the Ministry of Ecology and Natural Resources	A **National Commission (NC)** with 18 members representing parliament, the government, the private sector, research, and academic institutions. It is chaired by the Minister of Ecology and Natural Resources and the vice-chair is the Director of the State Hydrometeorological Service.	The State Hydrometeorological Service provided an office for the NC with two computers with access to the internet. The Climate Change Office Manager of the Ministry of Ecology and Natural Resources acts as the Secretary of the NC.
Montenegro	--	--	--
Kazakhstan*	--	*In process.* **Interagency Committee on Climate Change (IACCC)** approves CDM/JI projects. Consists of 12 ministries. Observer status to NGO, parliament, international donor and industry representatives.	The Climate Change Coordination Centre acts as the Executive Body of the IACCC.
Kyrgyzstan	National Climate Change Committee	A 2005 government decree established a **Climate Change Committee** chaired by the Deputy Minister of the Ministry of Nature Protection and Emergency Situations. A restructuring of the ministry into two (Ministry of Emergency Situations and the State Agency for Nature Protection and Forestry) put the DNA on hold	The **Climate Change Centre** (co-funded by UNDP and the Ministry of Nature Protection and Emergency) to act as the Secretariat.

* Bosnia and Herzegovina, Kazakhstan, Serbia and Tajikistan have not yet ratified the Kyoto Protocol.

Table 3. Cont.

Country	DNA Reported to the UNFCCC	DNA Legal Structure	DNA Administrative Unit/Secretariat
Serbia	Ministry of Environment[2]	*In process (proposed):* **DNA Supervisory Board (DSB)** will be formed by a representative from the Ministry of Science and Environmental Protection, Ministry of International Economic Relations, and Ministry of Mining and Energy. The Board will be chaired by the representative of the **Ministry of Science and Environmental Protection**	*In process:* It is proposed to establish a **DNA Technical Secretariat (DTS)** and **Inter-ministerial DNA Panel.** DTA will provide administrative and logistical support to DNA while the Panel will be responsible for providing opinions on general or specific sector-level issues. The IDP will be formed by a representative from each of the following institutions: Ministry of Science and Environmental Protection, Ministry of International Economic Relations, Ministry of Mining and Energy, Ministry of Agriculture, Forestry and Water Management, Ministry of Economy, Ministry of Capital Investment, the Energy Efficiency Agency, and the Environment Agency.
Tajikistan*	--	*In process.* **Ministry of Environmental Protection**	Secretariat to be housed within Ministry of Environmental Protection
Turkmenistan	--	*In process*	
Uzbekistan	Ministry of Economy	National CDM Board	Ministry of Economy

* Bosnia and Herzegovina, Kazakhstan, Serbia and Tajikistan have not yet ratified the Kyoto Protocol.
Source: UNDP BRC 2006

3. TRENDS IN INSTITUTIONAL DEVELOPMENT

Of the DNAs that have so far been created in South-Eastern Europe and the CIS, all are hosted by the government and all have designated the Ministry of Environment as the primary contact point for their DNA (except Uzbekistan where the DNA is hosted by the Ministry of Economy). Typically, an inter-ministerial board has been

---

[2] Serbia also reported the nomination of DNA to the UNFCCC Secretariat. However, in practice no formal authority was assigned to the contact person regarding CDM approval and the status if Kyoto Protocol ratification is still being debated in Serbia.

created that includes high-level officials from all the relevant functional ministries (energy, transport, agriculture, forestry, and economy). This body is usually headed by the Ministry of Environment and acts as the final decision maker on CDM project approval, requiring all, or a majority, of the ministries involved to agree on a project decision. A DNA secretariat, most often housed in the Ministry of Environment, then acts as the secretary for this body by coordinating all day-to-day activities, setting up meetings, and undertaking various outreach responsibilities.

In several cases, the DNA is hosted by an office that is specifically designated for climate change activities. This is for instance the case of Georgia, Armenia, and Albania. Georgia's DNA contact point is also the Director of the National Agency of Climate Change within the Ministry of Environment. This ensures that the DNA can draw on the eight staff members already assigned to climate change activities and can rely on their existing experience with other climate change activities undertaken as part of the UNFCCC. Alternatively, as in the case of Moldova, most of the work of the DNA is undertaken by the President and the Vice Chair of the National Commission (i.e., the DNA) responsible for implementing the Kyoto Protocol.

Other countries are considering placing the CDM evaluation function within a separate agency (e.g., Serbia). For example, Kyrgyzstan has nominated a Climate Change Centre to act as the DNA Secretariat. The Centre remains a government entity but is located outside the Environment Agency. The Climate Change Centre thus uses the experts and infrastructure developed for the UNDP/GEF Project on Climate Change which implemented the preparation of the 1st National Communication to UNFCCC.

It is important to realize that the official nomination of a DNA and notification of this to the UNFCCC Secretariat is only the first stage in the formal process of setting up the DNA. In order for the DNA to effectively undertake its functions, it should be granted full legal authority to review and approve CDM projects in accordance with the existing legal and regulatory framework of the host country. The legal provisions validating the DNA should contain clear statements regarding its legal justification, authority, objectives, organizational structure, functions, priorities and procedures.

In some cases, two or more legal documents will be required, with the first consisting of a notification of the location of the DNA and the second consisting of an agreement containing more specific information on the institutional set up, evaluation functions, and the sustainable development criteria of the unit. Armenia, for example, ratified the Kyoto Protocol in 2002, and later on, in 2003, the Ministry of Nature Protection was appointed as the DNA without having specific application procedures and criteria in place. In early 2005, the Ministry of Nature Protection issued its first approval letter for a CDM project. Moreover, in 2005, with the assistance of a EU-funded capacity-building project, Armenia started the process of developing and approving an institutional procedure for evaluating projects.

4. NATIONAL CDM PROJECT REVIEW AND APPROVAL PROCEDURES

Most countries in the region have proposed a two-step approval procedure, whereby project developers have the option of submitting a short Project Idea Note (PIN) for initial feedback, often in the form of a letter of endorsement, before the complete PDD is submitted for final approval (see Table 4).

Table 4. Sample of Proposal Review Cycles

Country	Initial Project Endorsement			Final Project Approval		
	Project Idea Note (PIN)	PIN Template	Time Frame	Project Design Document (PDD)	PDD Template	Time Frame
Georgia	Optional	Provided by DNA		Required	CDM EB's design	
Moldova	N/A	N/A	N/A	Required	Stage 1. Request authorization from government to negotiate project & sign PDD. Stage 2. Official approval by DNA (National Commission - NC): Baseline and monitoring methods must be approved by CDM EB before PDD in CDM format can be submitted to the NC.	A stage 2 decision must be made within 2 weeks. Rejected and revised proposals will also be examined within 2 weeks.
Armenia			10-15 working days	Required	CDM EB design	20 working days

Source: UNDP BRC 2006

The review and approval of the PDD and an assessment of its compliance with sustainable development goals and other host country requirements is normally conducted either by the DNA Secretariat itself or with the involvement of the inter-agency committee or a similar body. For example, in Georgia the CDM national board, an *ad hoc* Committee which includes high-level officials from sectoral ministries, has the final authority to make a decision on the approval or rejection of the proposed CDM activity. However, before this point, a technical screening and evaluation of the PDD will have been undertaken by the DNA Secretariat which is comprised of the UNFCCC National Focal Point and the department of Climate Change of the Ministry of Environmental Protection and Natural Resources of Georgia.

The Department of Climate Change already possesses sufficient technical facilities and human resources to facilitate this CDM review; experience which was gained during the UNDP/GEF-supported project on the preparation of the National Communication to the UNFCCC and other climate change-related capacity building projects. Several variations can be made to this process according to each individual country setting. For example, in Armenia, both the PDD and the draft approval/rejection letter are sent to appointed experts at the Ministry of Trade and Economic Development and the Ministry of Labor and Social Security to obtain their consent.

Some countries have set a time frame for when the project developer must be notified of a final decision. This helps to provide certainty for project proponents and investors. In the case of Moldova, the PDD review must take no more than two weeks after the official receipt of the proposal. A revised, initially rejected, proposal will also be reviewed within two weeks after resubmission. Armenia proposes to review PINs within 15 working days and PDDs within 20 working days. If additional information is requested on the PDD, the counting of the days will be halted until the required information has been received.

5. CONCLUSIONS

Regional assessment reveals that while significant progress has been made by South-Eastern European and CIS countries to establish required institutional frameworks for the Kyoto Protocol, some capacity gaps and barriers remain that limit these countries' participation in CDM. In particular, there is a great demand for external assistance to help CDM host countries in establishing DNA structures, setting up the institutional framework for evaluating projects, submitting letters of approval, defining sustainable development criteria, and examining project eligibility. Furthermore, many countries have expressed a need for support for issues related to the legal nature of CERs, taxation and the procedures for the transfer of CERs to foreign entities. All countries in the region would significantly benefit from targeted training to help relevant government experts conduct assessments of PINs/PDDs and Emission Reductions Purchase Agreements (ERPAs), as well as participating in specialized functions of the CDM Executive Board and other UN negotiations related to the flexible mechanisms. The training of ministry staff in the most recent developments in baseline and monitoring methodologies would be particularly helpful, especially if targeted towards the specific sectors and project types that are relevant for each country.

The Balkan countries are furthest behind in setting up DNAs and are in great need of support for initiating the interagency discussions related to deciding on an institutional set up and formulating the necessary decrees for doing so. Serbia and BiH in particular need to overcome the remaining suspicion towards the Kyoto Protocol in order to proceed with its ratification. Additionally, for the countries in the Western Balkan region that are preparing for future accession to the EU, special training would be useful for understanding the relationship between the EU's environmental finance schemes (i.e., renewable energy certificates and the EU Emissions Trading Scheme) and the mechanisms of the Kyoto Protocol.

In addition to differences in host-country capacity for CDM project formulation, review and approval, CDM projects, particularly concerning renewable energy, energy efficiency and biological mitigation/sequestration, are confronted with other significant barriers that deter private investment in these sectors such as:

- *Access to financing*: The high up-front costs of renewable and other climate-friendly technologies relative to projects dealing with industrial gases, combined with perceived risks associated with such technologies (e.g., risks in the supply of materials), deter investors. The current commodity model of the CDM market, whereby payments are made upon the delivery of Certified Emission Reductions (CERs), does not address this barrier;
- *Information and technological barriers*: There is limited access to information on technical options; most of the technologies are proven but not mature technologies;
- *Technical capacity*: There is weak capacity to design, construct, operate and maintain modern climate-friendly technologies;
- *Regulations/Licences*: These are unclear and often fragmented (e.g., waste treatment and waste disposal permits);
- *Ownership of CERs* can be a complex issue (for example, with afforestation/reforestation projects where the land on which CERS are generated is common property);
- *Carbon Process Risks*: There are greater uncertainties in assessing baselines and additionality, creating investment risks relating to CDM registration;
- *Long lead times*: In the case of afforestation/reforestation projects, carbon credits and revenue streams are displaced over a lengthy horizon;
- *Reversibility risk*: Sequestration projects face the additional risk that sequestered carbon can subsequently be lost (for example, through forest fires or disease). Expiring CERs (tCERs, lCERs) are designed to accommodate the reversibility risk at a systemic level, but they impair the attractiveness of sequestration projects for individual project developers and buyers.

Furthermore, the existing geographical and technology distributions of CDM projects globally and within South-Eastern Europe and the CIS reflect differences in the profitability of carbon investments. Carbon finance provides only an additional revenue stream that complements traditional revenue sources: it does not remove the need for traditional debt or equity financing to finance underlying projects. In many cases, the additional revenue stream created by carbon finance is unlikely to be sufficient for transforming a sustainable development-oriented CDM project into an attractive investment, should there be some fundamental problems with the economic feasibility of the underlying project. This is particularly relevant for projects aimed at energy efficiency improvement, the promotion of renewable energy (small-scale hydro, wind or solar) and bio-sequestration, which typically have strong sustainable development benefits but relatively low (or risky) carbon benefits.

As such, in order to redirect the flow of carbon financing towards projects and technologies with high sustainable development benefits and to expand the carbon market into new countries, there is a need to address the fundamental market barriers that limit the attractiveness of underlying projects to potential investors in general. This could call for the development of industry-specific policy instruments to complement the potential effect of carbon credits on the profitability of underlying projects. For example, preferential power purchasing arrangements for wind energy might need to be combined with carbon finance for the profitability uplift from carbon credits to prove sufficient to promote private investment in wind power. This approach would call for a concerted effort to remove the institutional, legal and capacity barriers to carbon finance as well as to combine a variety of different financing instruments such as ODA, local government support, carbon or private investment to boost the profitability of projects with high sustainable development impacts.

References

Climate Analysis Indicators Tool (CAIT) Version 3.0. 2006. Washington, DC: World Resources Institute. Available on-line: <http://cait.wri.org/cait.php> (last accessed April 2006).

UNEP Risø Centre on Energy, Climate and Sustainable Development (UNEP Riso). 2007. CDM Project Pipeline. Available on-line: <http://cdmpipeline.org/> (last accessed May 2007).

United Nations Development Programme/Bratislava Regional Center (UNDP BRC). 2006. '*How-to Guide: National Institutional Frameworks for the Kyoto Protocol Flexible Mechanisms in Eastern Europe and the Commonwealth of Independent States*', Bratislava, Slovakia: UNDP/BRC.

United Nations Development Programme (UNDP). 2006. '*The Clean Development Mechanism: an Assessment of Progress*', New York: UNDP.

United Nations Framework Convention on Climate Change (UNFCCC). 2007. Information and access to the different status of CDM project activities. Available on-line: <http://cdm.unfccc.int/Projects/index.html> (last accessed April 2007).

DEVELOPING COUNTRIES AND THE POST-KYOTO REGIME: BREAKING THE TRAGIC LOCK-IN OF WAITING FOR EACH OTHER'S STRATEGY

Joyeeta Gupta*

1. INTRODUCTION

As new evidence from the Inter-Governmental Panel on Climate Change[1] emerges to confirm the seriousness of the climate change problem, as new political happenings such as the publication and promotion of the Stern Report[2] and the Al Gore book[3] and film stimulate global discussion, as actual progress in dealing with the climate change problem stagnates, it becomes increasingly important to revisit the key sources of discontent in the regime for and with respect to developing countries and to explore new and innovative avenues for further cooperation. This paper identifies two sources of discontent and seeks ways to deal with them.

The first key source of discontent is the dampening, almost domino-like effect of the non-participation of the United States in the Kyoto Protocol of 1997.[4] I say almost domino-like effect since the climate change regime has not collapsed.[5]

---

* The author is Professor of climate change policy and law at the Institute for Environmental Studies of the Vrije Universiteit Amsterdam and is Professor of water law and policy at the UNESCO-IHE Institute for Water Education, Delft. This paper has been written as part of two projects – the Netherlands Organization of Scientific Research VIDI project on Intergovernmental and Private Environmental Governance: Sustainable Development, Good Governance and the Rule of Law (contract number: 452-02-031), and the European Commission-financed Adaptation and Mitigation (ADAM) Project (contract number: 98476).

[1] IPCC (2007). Climate Change Science: Fourth Assessment Report, forthcoming. See, for details, the other reports of the Intergovernmental Panel on Climate Change (IPCC) at <www.ipcc.org> which are published every five years. These reports present an assessment of existing peer-reviewed science on climate change. The IPCC was set up in 1988 by the WMO, UNEP and ICSU to assess climate change science.

[2] Stern Report: Stern, N., S. Peters, V. Bakhshi, A. Bowen, C. Cameron, S. Catovsky, D. Crane, S. Cruickshank, S. Dietz, N. Edmonson, S.-L. Garbett, L. Hamid, G. Hoffman, D. Ingram, B. Jones, N. Patmore, H. Radcliffe, R. Sathiyarajah, M. Stock, C. Taylor, T. Vernon, H. Wanjie, and D. Zenghelis. 2006. Stern Review: The Economics of Climate Change, HM Treasury, London.

[3] Gore, A. (2006). *An Inconvenient Truth, The Planetary Emergency of Global Warming and What We Can Do About It*, Rodale Press.

[4] Protocol to the Framework Convention on Climate Change (Kyoto), 37 *ILM* (1998) 22, in force 16 February 2005.

[5] Gupta, J. (2006). 'Environmental Multilateralism: Under Challenge?', in Edward Newman, Ramesh Thakur and John Tirman (eds.), *Multilateralism Under Challenge? Power, Normative Structure and World Order*, UNU Press, Tokyo, pp. 289-307.

W.Th. Douma et al., eds., The Kyoto Protocol and Beyond
© 2007, T·M·C·ASSER PRESS, *The Hague, The Netherlands and the Authors*

Although the US is not a party to the Kyoto Protocol, it is a party to the 1992 Climate Change Convention[6] and is thus engaged in the discussions. However, its non-participation has cast a damper on the enthusiasm of the remaining developed and developing countries to take action. A serious consequence of this is that rapidly developing countries like China and India may eventually choose to follow the US unilateral mode. If that happens then any serious attempt to mitigate global emissions will be postponed indefinitely. The likelihood of serious and far-reaching action by these two countries in the absence of US multilateral action within the Kyoto Protocol and its follow-up regimes is low. Hence, ways and means to ensure that these two countries can, like the US, actively (re)enter the space for negotiations becomes vital.

The second key source of discontent is the exposure to and increasing vulnerability of developing countries in general to the climate change problem and the growing difficulties they face in dealing with the problem of climate change (see section 2.3).

Hence, this chapter focuses on the following two questions: a) How can we keep the door for serious participation open for China and India, so that as and when it becomes politically feasible for the governments to participate effectively not only in terms of qualitative goals, but also in terms of quantitative targets, that this is also practically possible? b) What are the options that need further exploration in terms of enhancing North-South cooperation with respect to adaptation and emission reduction for all developing countries?

In order to address these two research questions, I will first present a general background to the problems facing developing countries (see section 2) and then move on to discuss the challenge of the giants (see section 3), before analyzing the complex problems facing the rest of the developing countries (see section 4).

2.   CHALLENGES FACING DEVELOPING COUNTRIES

2.1   Introduction

Let me first begin with a brief description of the main challenges facing developing countries within the climate change regime. These can be divided into two categories – the negotiation challenge and the challenge which the climate change problem itself poses to these countries.

2.2   The negotiation challenge

The negotiation challenge for developing countries has many dimensions. Moving from the broader to the more specific picture, one can argue that there are seven types of challenges.

---

[6] United Nations Framework Convention on Climate Change, (New York) 9 May 1992, in force on 24 March 1994; 31 *ILM* 1992.

First, the global governance set up for the 21st century embodies many competing elements. These include the contradictory ideological messages communicated by different actors and different regimes within the system. While the open trade and liberalization message is marketed, *inter alia*, through the World Trade Organization and the Banks tend to focus on allowing the market to function freely and enhance private sector participation in resource management, the environmental agreements tend to focus on restraining production and consumption processes in order to promote sustainable development.[7]

Second, as we move to climate change-related treaties and regimes, we find a number of different agreements and organizations, each with its own logic and its own rules and regulations. These different treaties and bodies deal with biodiversity, deforestation, desertification, water, wetlands, depletion of the ozone layer and have sometimes synergetic and often conflicting goals.[8]

Third, as we zero in on the climate change regime we find, on the one hand, the multilateral climate change negotiations and, on the other, the numerous unilaterally driven negotiations on a number of climate change-related issues initiated by the US which tend to create an enormous amount of confusion in the process.[9]

Fourth, within the multilateral climate change regime, the classification of countries as developed (Annex I) and developing (non-Annex I) is not really conducive for healthy North-South negotiations. The Annex I list includes poorer countries, the non-Annex I countries include richer countries and this creates problems in the negotiating process. [10]

Fifth, there are critical problems associated with the formal process of negotiations. Negotiations are often held in multiple non-plenary sessions, where the working language is English, which often continue past the time set for the end of the meetings, and often tax the limited developing country participants per country through very long negotiating sessions.[11] Apart from these more practical challenges, there are two more serious legal issues. The first is that as environmental law develops rules not based on state practice but on negotiations based on peer reviewed, primarily Anglo-American science, it tends to develop solutions that of necessity will be alien to developing country cultures and practices.[12] The second is that as not only

---

[7] Gupta, J. (2006). 'The European Union and Climate Change: Challenges and Options', in Peeters, M. and Deketelaere, K. (eds.), *EU Climate Change Policy: The Challenge of New Regulatory Initiatives*, Edward Elgar, London, pp. 297-315.

[8] Asselt, H. van, Gupta, J. and F. Biermann (2005). 'Advancing the Climate Agenda: Exploiting Material and Institutional Linkages to Develop a Menu of Policy Options', *Review of European Community and International Environmental Law*, Vol. 14(3), pp. 255-264.

[9] Gupta, J. (2005). *Who's Afraid of Global Warming?,* Inaugural Address as Professor of Climate Change: Policy and Law, Vrije Universiteit Amsterdam, 21 October, ISBN 90-90201-43-2.

[10] Gupta, J. (2003). 'Engaging Developing Countries in Climate Change: (KISS and Make-Up!)', in David Michel (ed.), *Climate Policy for the 21st Century: Meeting the Long-Term Challenge of Global Warming*, Centre for Transatlantic Relations, Johns Hopkins University, Washington D.C. pp. 233-264.

[11] See n. 9 *supra*.

[12] Gupta, J. (2006). 'Industrial Transformation and International Law', in X. Olsthoorn and A. Wiezorek (eds.), *Sciences for Industrial Transformation: Views from Different Disciplines*, Kluwer Academic Publishers, pp. 53-74.

treaties and protocols develop rules, but also the Conference of the Parties, we are increasingly dispensing with the notion of state consent for critical issues. While many developing country legal purists will argue that only the Convention and Protocol are important, it is increasingly being recognized that the bulk of the decisions are being taken at the Conference of the Parties. To the extent that such decisions are only being prepared for treaty negotiation – it is often too late to change these decisions during such final negotiations if countries have not been actively engaged in the preparatory processes.[13]

Sixth, the process of forming coalitions within the developing world, although increasingly becoming more institutionalized, lacks content and organization. The lack of resources to develop common inter-sessional positions based on well-defined scientific papers that articulate national and regional interests is a major stumbling block for these countries.[14]

Seventh, the preparatory process within developing countries is limited as a result of their ideological vacillation, their relative lack of scientific information that is directly pertinent to the negotiations, the lack of public and political interest in climate change-related issues, and the inability to go beyond simple, rhetorical demands as to how the climate change problem can be addressed. This tends to lead to a hollow mandate for negotiating purposes, although clearly there are exceptions.[15]

2.3   The climate challenge

In addition to the more practical challenges developing countries face with respect to the climate change negotiations, they face the real and pressing danger of the potential impacts of climate change, and the need to modernize their infrastructure and systems so as to minimize their contribution to greenhouse gases, and the risk of having to close down new operations as and when climate change obligations become binding on these countries.

The developing countries have submitted two types of documents to the climate change secretariat indicating the kinds of impacts they expect to face and the kinds of climate change-related challenges that are serious for them. One of these is the National Communications which describe, *inter alia*, national emission inventories and policies and the other is the National Adaptation Plan of Action.

The National Communications from more than 120 developing countries reveal that significant numbers of these countries have very large populations living below the poverty line. This has two implications with respect to climate change.

---

[13] Brunnée, J. (2002). 'COPing with Consent: Law-Making Under Multilateral Environmental Agreements', *Leiden Journal of International Law*, 15(1), pp. 1-52; see also n. 9 *supra*.

[14] Gupta, J. (2002). 'The Climate Convention: Can a Divided World Unite?', in Briden, J. and Thomas, E. Downing (eds.), *Managing the Earth: The Eleventh Linacre Lectures*, Oxford University Press, Oxford, pp. 129-156.

[15] Gupta, J. (1997). *The Climate Change Convention and Developing Countries - From Conflict to Consensus?*, Environment and Policy Series, Kluwer Academic Publishers, Dordrecht, p. 256.

First, that these countries prioritize, at least rhetorically, poverty reduction; and, second, that the poorer population is more likely to be less resilient to the potential impacts of climate change.

The National Communications cover tiny countries like Niue which has a population of 3,000 people and large countries like China which has a population of 1.2 billion. For most countries the priority issues were poverty reduction, water and food security. Most of these countries have unstable governments and are in the process of transition towards becoming market economies and face problems as a result of fluctuations in the market and currency sector. For different countries, different sources of greenhouse gases are important. For Africa, the key gas appears to be methane; while for Asia and Latin America, CO_2 emissions are critical. For most parties, greenhouse gases were mostly emitted by the energy sector. For 45 countries, the key sector is agriculture, and for six it is waste. Land use change is a net sink in Asia and Africa, but is a source in Latin America.[16] Emission per capita in Africa was about 2.4 tonnes, while some of the Central European and Asian countries had emission levels of about 5.1 tonnes.

Many of these countries are taking measures to deal with these gases for domestic and international reasons. Some attention is focused on energy supply and demand; but explicit focus on fuel-switching is low. For most of these countries, there is a considerable fear that changing climatic patterns would affect food and water security as well as their coastal and marine resources.[17]

While some of the larger countries like China, India and Brazil have taken a large number of measures, these countries are very cautious about how they present such measures to the international community arguing that the primary responsibility for action does not rest with them. Nevertheless, they are quick to point out a number of measures taken – such as Brazil's National Alcohol Programme, China's Greenlight Programme and India's renewable energy programme.[18] China emphasizes its efforts to decouple energy consumption from economic growth and India submits that it is consciously factoring its commitment to the climate change convention in its national policy.[19]

While most countries elaborate on the possible measures they can take, they also emphasize the lack of resources. Kazakhstan estimates that it needs USD 5 billion to reduce its emissions; Djibouti – 88 million to develop geothermal and ocean energy, and Haiti claims 300 million is necessary to develop hydropower to replace diesel generators.[20] These Communications reveal the current state of na-

---

[16] FCCC/SBI/2005/18; p. 9.

[17] Ibid., p. 15.

[18] Sixth Compilation and Synthesis of Initial National Communications from Parties Not Included in Annex I to the Convention, Addendum on Sustainable Development and the Integration of climate change concerns into medium- and long-term planning, FCCC/SBI/2005/18/Add.1, 25 October 2005.

[19] India's Initial National Communication to the United Nations Framework Convention on Climate Change, Ministry of Environment and Forests, Government of India, 2004; p. xiii.

[20] FCCC/SBI/2005/18/Add.3, p. 17.

tional awareness of their contribution to the climate change problem, what countries can do and the level of resources they see as needed. These Communications should help to provide a common knowledge base for these countries to negotiate more effectively in the future. However, they also reveal how general and vague some of the information is.

The NAPAs (National Adaptation Programmes of Action) are programmes to assist the 50 least developed countries to prepare an analysis of the impacts of climate change and how they can best adapt on a priority basis to the potential impacts. The recognition of the special needs of LDCs (UNFCCC Article 4.9) and decisions 5/CP.7 and 28/CP.7 of the Conference of the Parties led to the idea of the NAPAs. These documents identify the strategies of local populations to deal with climatic variation and seek to find ways to make national priorities on the basis of this. The NAPAs are illustrative of the kinds of problems developing countries expect to face. The NAPA of Comoros prepared in 2006, highlights the extreme vulnerability of the three islands of Comoros. Since the agricultural sector employs 80% of the population and contributes about 40% of the income, 'the [potential climate change] impacts could undermine the efforts undertaken by the country to fight poverty' (p. 31). The NAPA lists 13 priority areas of action including the introduction of crop varieties that are more adapted to drought; the defence and restoration of degraded soils; the reconstitution of basin slopes; an increase in water supply and the improvement of water quality; the need to fight malaria through eliminating larva and distributing mosquito nets; the use of local non-metallic materials for the construction of low-priced housing; fodder production for goat breeding; provender production (for chicken breeding); the introduction of Fish Concentration Mechanisms; short-term conservation of fish under ice to reduce losses after catches, due to high temperatures; the setting up of an early warning and surveillance system on situations of climatic risks (cyclones, floods, droughts); and support to eye, medical and surgical care to combat communicable diseases and eye affections (p. 45). This list communicates that some degree of thought has been put into the process of prioritizing national problems and each has a price tag ranging from 75,000 USD to over a million. The NAPA focuses on poverty alleviation and concludes that the barriers to implementation include limited knowledge and awareness among the local communities and developmental stakeholders; the unpredictability of climate change effects; institutional weaknesses; difficulties in obtaining the necessary financial resources; and concerns over the daily life of the most vulnerable populations. The lack of a policy on climate change is expected to be remedied through the installation of a national commission for adaptation to climate change. The democratic government elected in 2006 is hopeful about being able to address poverty reduction and corruption.[21]

---

[21] Comoros (2006). National Adaptation Programme of Action, Ministry of Rural Development, Fisheries, Handicrafts and the Environment, Union of Comoros. <http://unfccc.int/resource/docs/napa/com01e.pdf>.

Malawi submitted its NAPA in 2006.[22] With a life expectancy of 34 years primarily as a result of vulnerability to diseases including HIV/Aids and malaria, Malawians have a peaceful and democratic society and are working towards establishing good governance. Around 85% of the population is dependent on agriculture. The NAPA sees climate change as impacting on water, food, energy and health security. The country is rain-fed and mostly dependent on a single crop of maize. The report focuses on water management including water harvesting, water conservation, small-scale irrigation, and access to water and water treatment. Food security can be enhanced through crop diversification, seed storage, fish gene banks, fish farming sites, and food banks. Afforestation along riparian boundaries can provide income as well as preventing siltation and pollution. To cope with flooding the report suggests the establishment of early warning systems and monitoring and the construction of dams, levees, dykes and canals. While the report demonstrates that Malawi is seriously concerned about the impacts of climate change, it focuses more on advanced technology and the empowerment of local people through education and climate awareness.

These two reports provide an idea of the kind of information that the 17 NAPAs[23] that have been presented to the UNFCCC secretariat reveal.

2.4 Inferences

The above text illustrates that the developing countries face a number of serious negotiating problems. It also shows that there is growing awareness of how serious the climate change problem is for their own countries and how it links up with many of their national priorities.

3. THE CASE OF CHINA AND INDIA: KEEPING THE DOORS FOR ENTRY OPEN

3.1 Introduction

China and India have been participating actively in the climate change negotiations since the start in 1990. However, they have been clear in stating at the international conferences and through their National Communications that the main responsibility for addressing the climate change problem rests with the developed countries. This has once more been reiterated by China at the G8 Conference in Germany in June 2007.

---

[22] Malawi (2006). National Adaptation Programme of Action, Malawi Ministry of Mines, Natural Resources and Environment: Environmental Affairs Department (March 2006). Malawi's National Adaptation Programme of Action (NAPA) Under the United Nations Framework Convention on Climate Change (UNFCCC), Lilongwe, Malawi. Link: <http://unfccc.int/resource/docs/napa/mwi01.pdf>.

[23] The other NAPAs are from Bangladesh, Bhutan, Burundi, Cambodia, Djibouti, Eritrea, Haiti, Kiribati, Lesotho, Mauritania, Madagascar, Niger, Rwanda, Samoa and Senegal. <http://unfccc.int/national_reports/napa/items/2719.php> (last visited on 8 June 2007).

My personal conviction is that, although this is difficult to prove, had the US participated in the Kyoto Protocol, it would only have been a matter of time before these two countries had made some kind of formal and quantitative commitment to deal with the climate change issue. However, since the US has not done so, and instead there are a number of bilateral and unilateral-driven agreements being negotiated simultaneously by the US, the question is the following: in which direction will these two countries go?

This is a critical question since both these countries are presently experiencing rapid economic growth and it is more than likely that their greenhouse gas emissions will increase accordingly. Some expect that the total emissions of China are likely to exceed that of the US by 2009. Any effort by the rest of the world to restrain its own growth will be ultimately insignificant if the US, China and India stay outside the purview of future quantitative targets adopted by the climate change regime. This section therefore argues that it is of vital interest to try and push a number of different policy options in different arenas to ensure that when the political moment arrives, it is not too complex for the US to re-enter the regime and for China and India to engage constructively in discussions regarding their greenhouse gas emissions.

This section proposes three ideas for keeping such doors open. The first is mainstreaming within the UN, the second is actively engaging local government in the process, and the third is putting pressure on the government of the US through litigation; an option that eventually also becomes of critical importance in a country like India.

3.2 Mainstreaming in the UN

Mainstreaming within the United Nations (UN) would imply integrating climate change into all relevant bodies within the UN. Climate change is closely related to energy, agriculture, industry, the environment, forestry, trade, housing, health and water, and possibly many other sectors. The possible impacts of climate change include rises in the sea level, changes in precipitation patterns and melting glaciers. These will affect water supplies and local climates and hence agriculture, infrastructure and housing, industry, trade patterns, ecosystems and tourism. These subjects are dealt with via a variety of different bodies in the UN – namely the International Energy Agency and the International Atomic Energy Agency, the Food and Agriculture Organization, UNCTAD, the United Nations Environment Programme, the United Nations Forum on Forests, the World Trade Organization, the International Maritime Organization, Habitat, the World Health Organization and the 23 organizations of the UN that work on water-related issues. Mainstreaming would involve ensuring that the sources of greenhouse gases and the potential impacts of climate change are taken into account in the relevant policies of each of these different bodies. Several publications explore these issue linkages,[24] some

---

[24] See, for example, Bradnee Chambers, W. (ed.) (2001). *Inter-linkages: The Kyoto Protocol and the International Trade and Investment Regimes*, United Nations University Press, Tokyo.

bodies exist to promote cooperation, but much more can be done to package the message consistently in different policy spheres. Ensuring that environmental issues are taken into account in the bilateral and plurilateral investment treaties is also important, as this is presently not the case.[25] Enhancing cooperation within the different policy arenas is complicated as each of these has a different logic, different parties and different histories.[26] In terms of the trade regime, it is vital that a balance is actively sought between helping countries exploit their comparative advantages and between seeking to promote sustainable use and production of resources.

Such mainstreaming is important as it ensures that a consistent policy message is sent to all countries in the world. It is also important because it clearly links environmental problems with development problems, the latter are clearly seen as current pressing priorities in the developing world.

3.3 Encouraging local government policy

Especially where central governments are reluctant to take action on climate change, but even where central governments wish to take action, there is room for local initiative. Efforts to support such local initiative may be a key way to promote a domestic constituency on climate change.

At present we are seeing a number of initiatives at the local level worldwide aimed at increasing the potential for action on climate change in these countries. This is partly being motivated by irritation at the lack of central government action in the area of climate change and partly because there is a fear that central government action does not go far enough and much more needs to be done, and that impacts will be felt primarily locally within cities and villages. Many scientists also argue that policies will only be effective if they are taken at local level and this is another argument to push for measures at this level.[27]

The International Coalition of Local Environmental Initiatives has been promoting sustainable development at local level all over the world and 539 cities in 68 countries have joined the programme. Some 122 municipalities in Canada,[28] 75 local governments in Europe,[29] 25 cities in Latin America,[30] 37 in Sub/Saharan

---

[25] See <http://www.worldbank.org/icsid/treaties/treaties.htm> for a list of the treaties. See also Peterson 2004; and Cosbey, et al. 2004.

[26] See Oberthür, S. (2002). Clustering of Multilateral Environmental Agreements: Potentials and Limitations, 2(4), pp. 317-340.

[27] See the forthcoming special issue on action in local governments in *Local Environment* and the forthcoming special issue on Climate Change: National and Local Policy Opportunities in *Environmental Sciences* and Bulkeley, H. and M. Betsill (2003). *Cities and Climate Change: Urban Sustainability and Global Environmental Governance*, New York: Routledge.

[28] Alberta(11), British Columbia(39), Manitoba(8), New Brunswick(12), Newfoundland(5), North West Territories(3), Nova Scotia(6), Nunavut(1), Ontario(32), Prince Edward Island(1), Quebec(5), Saskatchewan(2), Yukon(1).

[29] <http://www.iclei-europe.org/index.php?id=climate_and_air>.

[30] <http://www.iclei.org/index.php?id=528>.

Africa,[31] and 17 in India[32] have joined the coalition. The enthusiasm is even greater in countries that have not ratified the Kyoto Protocol such as Australia where 200 municipalities have joined up[33] and 89 in the US. On the 21st of February 2007, Delhi became the 500th member of ICLEI! In Australia, it is reported that municipalities participating in climate change programmes have reduced 2.9 million tonnes of carbon dioxide-equivalent (CO_2e).[34]

Where central governments have a number of geopolitical and strategic issues they need to take account of, city and provincial governments have more freedom. In the US, where some provinces have higher emissions than small developed countries,[35] action taken at this level can be highly effective. New York State has adopted a target to reduce its greenhouse gasses by 5% by 2010 and 10% by 2020 with respect to 1990 levels. Connecticut, Maine, Massachusetts, New Hampshire, Rhode Island, Vermont, Newfoundland, New Brunswick, Nova Scotia, Prince Edward Island and Quebec have a plan to reduce their emissions by 10% by 2020 with respect to 1990 levels. California has taken a lead role by aiming to reduce its greenhouse gas emissions to 2000 levels by 2010, to 1990 levels by 2020 and to reduce by 80% below 1990 levels by 2050.[36]

In China, the Central Government has allocated targets to lower governments. Jilin Province has to reduce its emissions by 30% while Tibet has to reduce its emissions by 12%.[37] The Central Government is trying to engage lower governments to actively think about their potential to reduce emissions. However, a key area of interest is Beijing. Beijing, with a population of more than 15 million and a per capita income of USD 6000 at present, is highly motivated to clean up its local pollution in time to make a good impression and meet the WHO standards as the host of the Olympics in 2008. As a result, the city has actively started developing policies on fuel-switching, energy consumption in the building sector (e.g., aiming at a 65% energy savings target in new public buildings by 2010), energy consumption and vehicular emissions in the transport sector (e.g., by making the Euro IV standard for cars compulsory since January 2007), and is providing subsidies for green lighting, renovating coal-fired boilers, and for a fuel switch in public transport to compressed natural gas. The relatively high local revenue of Beijing makes it possible to invest heavily in renewable energy and to provide subsidies to public

---

[31] <http://www.iclei.org/index.php?id=700>.

[32] <http://www.iclei.org/index.php?id=1089>.

[33] <http://ccp.iclei.org/ccp-au/index.cfm>; for information on the US, see 4.3.5.

[34] Press Release, dated 8/12/2006.

[35] New York, Texas, California, Ohio, and Pennsylvania have higher emissions than the Netherlands, Belgium, Austria and Denmark. See Centre for Clean Air Policy, *State and Local Climate Change Policy Actions* (Centre for Clean Air Policy, 11 October 2002), at 1.

[36] On 1 June 2005, Governor Schwarzenegger of California signed Executive Order # S-3-05 which lists these targets; see for more details the website of the government in California on the issue of climate change <http://www.climatechange.ca.gov/climate_action_team/>.

[37] F. Teng and A. Gu (2006). 'Climate Change: National and Local Policy Opportunities in China', paper presented at a Climate Change Course, EU Asia Link Project, Amsterdam, January 2007.

transportation. By linking up climate change policy with air pollution policy and the Olympic Games, Beijing can make an impressive reduction in greenhouse gas emissions with possible spill-over impacts in neighboring states and on the central government.[38]

In August 2001, the Cities for Climate Protection programme was initiated in India and seven municipalities joined. The first phase was seen as successful and another nine are participating in the second phase. These cities are helped by ICLEI to prepare emission inventories, identify policy measures and share experiences. ICLEI also helps to find funders to support projects in these cities. At present nine projects are being stimulated through this process in different cities.[39] New Delhi, which became the 500th member of ICLEI, can be seen as a microcosm of India with its own very rich and very poor people. As the Commonwealth Games scheduled for 2010 in New Delhi approaches, Delhi wishes to present itself as a modern and conscious city. The chief minister of the city explains that in collaboration with civil society and the Supreme Court in the city, there is enough momentum to clean up the city – by phasing in compressed natural gas in public transport, cleaning up the local habitat and environment, promoting environmental consciousness in schools, promoting energy efficient lighting in government offices and beyond, improved street lighting and a number of other measures. However, with an influx of half a million people per year the city has to constantly run to maintain itself. The motivation to address environmental issues in Delhi may be one reason to target lower governments as a possibly more effective partner in climate change policy.[40]

Promoting policy change at local level can reveal the room for policy action that actually exists at these levels and international collaboration that nurtures such developments at sub-national level will help to create broad-based support for long-term measures.

3.4 Litigation

Another possible avenue for promoting action is to pursue litigation. The potential for litigation options is being actively explored in the literature.[41] At the interna-

---

[38] Teng, Fei and Alun Gu (forthcoming). 'Climate Change: National and Local Policy Opportunities in China', *Environmental Sciences*.

[39] <http://www.iclei.org/index.php?id=1089>.

[40] Based on an interview with the Chief Minister of Delhi, Mrs. Shiela Dikshit, on 25 April 2007.

[41] Bradford C. Mank, 'Standing and Global Warming: Is Injury to All, Injury to None?', 35 *Environmental Law*, 1 (2005). Eduardo M. Penalver, 'Acts of God or Toxic Torts? Applying Tort Principles to the Problem of Climate Change', 38 *Nat. Resources J.* 563 (1998). David A. Grossman, 'Warming up to a Not-so-Radical Idea: Tort Based Climate Change Litigation', 28 *Colum. J. Envtl. L.* 1 (2003). Myles Allen, 'Liability for Climate Change: Will it ever be possible to Sue Anyone for Damaging the Climate?', Commentary in 421 *Nature* 891 (2003)892. Alexander Gillespie, 'Small Island States in the Face of Climate Change: The End of the Line in International Environmental Responsibility', 22 *UCLA Journal of Environmental Law and Policy* 107 (2004), Elizabeth E. Hancock, Red Dawn, Blue Thunder, 'Purple Rain: Corporate Risk of Liability for Global Climate Change and the SEC Disclosure Dilemma', 17 *Geo.Int'l Envtl.L.Rev.* 223 (2005), Rebecca E.

tional level, a court case has been submitted by a representative of the Inuit people to the Inter American Commission on Human Rights.[42] Some authors have examined if the small island states and specific developing countries could seek redress from the International Court of Justice,[43] or before the Tribunal of the World Trade Organization using the argument that US failure to ratify the Kyoto Protocol amounts to an illegal subsidy to US companies.[44]

Unlike international litigation options that are relatively difficult to pursue, litigate and enforce, there is some potential in domestic litigation. Litigation is being actively pursued as an option in the US and Australia.[45] Such litigation can cover a variety of legal grounds.[46]

Court cases in developing countries include a case initiated in 2005 in Nigeria arguing against the legality of gas flaring leading to 70 million tonnes of CO_2 emissions annually.[47] The initial judgment called for halting the gas flaring[48] but appeals are in place. In Argentina, a court case was initiated against the government for not implementing Article 6 of the Climate Convention and providing information in the aftermath of the Sante Fe floods.[49]

Jacobs, 'Treading Deep Waters: Substantive Law Issues in Tuvalu's Threat to Sue the United States in the International Court of Justice', 14 *Pacific Rim Law and Policy Journal* 103 (2005), Angela Lipanovich, 'Smoke before Oil: Modelling a Suit Against the Auto and Oil Industry on the Tobacco Tort Litigation is Feasible', 35 *Golden Gate U.L.Rev.* 429 (2005), Kristin L. Marburg, 'Combating the Impacts of Global Warming: A Novel Legal Strategy', *Colo.J.Int'l Envtl.L.& Pol'y 2001 Yearbook*, 171 (2002), William C.G. Burns, 'The Exigencies that Drive Potential Causes of Action for Climate Change Damages at the International Level', 98 *ASIL PROC* 223(2004); Richard W. Thackeray, 'Struggling for Air: The Kyoto Protocol, Citizens' Suits Under the Clean Air Act, and the United States Options for Addressing Global Climate Change', 14 *Indiana International and Comparative Law Review* 855 (2004), Roda Verheyen, 'Climate Change Damage in International Law', Dissertation zur Erlangung des Dr. iur Universitat Hamburg, Fachbereich Rechtswissenschaft (2003).

[42] Petition to the Inter American Commission on Human Rights Seeking Relief from Violations Resulting from Global Warming Caused by Acts and Omissions of the United States (7 December 2005).

[43] Gillespie, *supra* n. 41. Jacobs, *supra* n. 41. Verheyen, *supra* n. 41.

[44] Burns, *supra* n.41; see also Meinhard Doelle, 'Climate Change and the WTO: Opportunities to Motivate State Action on Climate Change through the World Trade Organization', 13 *RECIEL* 85 (2004).

[45] *Australian Conservation Foundation* v. *Minister of Planning* 2004 VCAT 2029 (29 October 2004); <www.austlii.edu.au/au/cases/vic/VCAT/2004/2029.html>. See also <www.climatelaw.org/media/CANA.Australia>; <www.austlii.edu.au/au/cases/vic/VCAT/2004/2029.html>; <http://www.climatelaw.org/media/Australia.emissions.suit>.

[46] For details, see Gupta, J. (2007). 'Legal Steps Outside the Climate Convention: Litigation', *Review of European Community and International Environmental Law*, 16(1) 76-86.

[47] *Barr. Ikechukwu Okpara, et al.* v. *Shell Petroleum Development Company of Nigeria Limited, et al.* (Suit No. FHC/CS/B/126/2005, 20 June 2005), filed in the Federal High Court of Nigeria, in the Benin Judicial Division.

[48] *Jonah Gbemre* v. *Shell Petroleum Development Company Nigeria Ltd*; Nigerian National Petroleum Corporation and Attorney General of the Federation, Decision of the Federal High Court of Nigeria in the Benin Judicial Division, Benin City (Suit No, FHC/B/CS/53/05, 14 November 2005).

[49] See also R. Verheyen, 'Climate Change in Courts World-Wide', Paper for the Conference 'Kyoto Plus - Escaping the Climate Trap', 28/29 September, 2006, organized by Heinrich Boll Stiftung, Wuppertal Institute for Climate, Environment and Energy, WWF and European Climate Forum.

Although climate change litigation has not begun in India, public interest litigation on environmental issues has been quite successful. Some cases have led to a fuel-switch in public transport in Delhi with impacts around the capital, the phasing out of older commercial vehicles and new standards for cars, leading to a reduced rate of growth of greenhouse gas emissions.[50] Domestic litigation can be the driving force in the US and India, but whether it has similar potential in China remains to be seen.

4.   VULNERABILITY OF DEVELOPING COUNTRIES (2000)

4.1   Introduction

Having discussed the three large entities, it is important to look at the remaining 148 non-Annex I countries. For a significant number of these countries, the key issue is adaptation. For some of these countries, the transfer of technology to promote a shift towards more energy efficient situations is also important. This section briefly discusses the situation with respect to aid to developing countries on climate change, and the mechanisms to promote technology transfer.

4.2   Aid

I would like to make three points with respect to aid, first on the sensitivities of linking aid to climate change, second that lessons learnt in aid history should be taken into account in cooperative policy on climate change, and third – to briefly recapitulate the current state of assistance on climate change.

Aid and climate change has always been a sensitive issue for developing countries. From the start the focus has been that assistance for climate change should come over and above the assistance that has been made available to developing countries as part of development cooperation. This was the famous call for 'new and additional' resources to deal with climate change which was included in the text of the Convention.[51] However, the development cooperation budget itself has remained rather low and the financial assistance provided to developing countries on climate change has not increased significantly.

Lessons from fifty years of aid delivery should be incorporated into climate change aid policy. These fifty years reveal that aid works when aid-related challenges in the developed countries are addressed. These include the need to move away from disbursing committed funds to *ex ante* designated recipients to a two-

---

[50] S.C. Writ Pet. (Civil), *M.C. Mehta* v. *Union of India* (29 July 1998), (No. 13029/1985), AIR 1998 SC 2963, found at <http://www.elaw.org/resources/text.asp?ID=1051>; S.C. Writ Pet. (Civil), *M.C. Mehta* v. *Union of India* (5April 2002), No. 13029/1985, AIR 2002 SC 1696 found at <http://www.elaw.org/resources/text.asp?ID=1102>.

[51] See Art. 4.3 of UNFCCC, *supra* n. 6.

way dialogue focusing on joint interests;[52] to ensure that donors have good contextual knowledge of where the aid is to be delivered;[53] and that aid is not tied which is both more expensive and less productive.[54] The problems in relation to the recipient countries are poor governance, the substitution effect of aid,[55] and the impacts of aid delivery on policy substitution where recipients leave certain policy areas to aid agencies.[56] The problems in the donor-recipient relationship include the possible mismatch between priorities of both parties which is then attempted to be resolved through conditionality; the poor diagnosis of the contextual aspects through over-reliance on theoretical solutions and hence the poor implementation of these tools; the realization that technical assistance is often too expensive, inappropriate and fosters dependency;[57] the mismatch between partners at the human level and the enormous administrative burden that aid accountability has put on some developing countries. An average African country has to prepare 10,000 quarterly reports to the different donors to account for the money.[58]

In 2005, the OECD member states provided development assistance amounting to USD 106.8 billion, about 0.33% of their joint Gross National Income. A large percentage of this money is earmarked for development assistance, but there are increasing discussions on using these resources for environmental assistance and for mainstreaming environmental issues into development assistance as the EU is at present aiming to do.

The Least Developed Countries Fund, established in 2001,[59] had, as of mid 2006, USD 41.8 million in cash, promissory notes and investment income. The fund finances the compiling of the National Adaptation Programmes of Action (NAPA) which are presently taking place and intends to fund country-identified adaptation policies. Around 50 countries are classified by the UN as LDCs.

The Special Climate Change Fund was established also in 2001 and has a budget of USD 36.7 million. This Fund will finance adaptation projects but is not restricted to that. At present seven projects are in the pipeline. These include projects in Ecuador; the Andean region; India; the Pacific Islands; Tanzania; in Fiji, the

---

[52] Sevensson, J. (2003). 'Why Conditional Aid Does Not Work and What Can be Done About it?', *Journal of Development Economics* 70: 381-402 at p. 383; Alseina, A. and B. Weder (2002). 'Do Corrupt Governments Receive Less Foreign Aid?', *The American Economic Review*, 92(4): 1126-1137. Burnside, C. and D. Dollar (1997). 'Aid, Policies and Growth', World Bank, Policy Research Working Paper 1777, The World Bank, Washington, D.C.

[53] Pronk, J.P. (2001). 'Aid as a Catalyst', *Development and Change* 32; 611-629.

[54] Pronk, J.P. (2003). 'Aid as a Catalyst: A Rejoinder', *Development and Change* 343(3): 383-400.

[55] Ndulu, B.J. (2002). 'Partnerships, Inclusiveness and Aid Effectiveness in Africa', in FitzGerald, V. (ed.), *Social Institutions and Economic Development: A Tribute to Kurt Martin*, Kluwer Academic Publishers, 143-168; Jepma, C.K. (1991). 'The Tying of Aid', OECD, Paris.

[56] Sevensson, J. (2000). 'When is Foreign Aid Policy Credible? Aid Dependence and Conditionality', *Journal of Development Economics* 61: 61-84.

[57] Joint European NGO report (2006). 'EU Aid: Genuine Leadership or Misleading Figures? An Independent Analysis of European Governments' Aid Levels'. Brussels: Concord.

[58] Idem.

[59] UNFCCC Decisions 7/CP.7.

Maldives and the Solomon Islands; and one project focuses on health issues at the global level.

The Adaptation Fund announced in the Kyoto Protocol and established in 2001 is dependent on the levy on the Clean Development Mechanism (CDM). The more successful the CDM, the more money will be raised for adaptation. At present negotiations are taking place regarding the modalities and priority setting for this fund.

4.3   Mechanisms to promote technology transfer

The five mechanisms to promote climate-relevant technology transfer to the developing countries include the Global Environment Facility, Activities Implemented Jointly, the Clean Development Mechanism, technology transfer and the Special Climate Change Fund.

The Global Environment Facility, launched in 1991, and established as the financial mechanism for the climate convention in 1992, provides incremental funding for the global benefits of projects in the developing countries. A recent evaluation of the GEF reveals[60] that one third of the total GEF funds (USD1.63 million) have been allocated to climate change projects. Only 43 of the 207 projects had been completed as of 2004. In terms of catalyzing change in countries, energy efficiency projects have been the most effective while renewable energy projects have been the least effective. While all eligible countries have received funding, the highest emitters in gross terms have received the most. As of 2004, the funding mostly went to China, India, Mexico, Brazil, the Philippines, Poland, Morocco, Uganda, Tunisia and India. Asia received the most, Sub-Saharan Africa the least. However, investments in these countries did not reflect the needs expressed in National Communications.

Activities Implemented Jointly, launched in 2005, renewed in 1999, 2002 and 2004, and 2006,[61] promotes a pilot phase of projects in which developed country investors invest in developing countries and countries with economies in transition in return for emission reductions that are calculated but not credited to the investor country. The sixth synthesis report of the Subsidiary Body on Scientific and Technological Advice reported that as of 2006, 157 AIJ projects had been registered in 11 developing countries and the rest are in the developed world. Of the 71 projects in developing countries, 57% are in Latin America, 26% in Asia and the Pacific, and 17% in Africa. Sweden and the US are the largest investors. While 22% of the projects are directed at the energy sector, 43% of the expected emission reduction will come from fugitive gas capture and 34% from forestry.[62] Since the US has submitted a letter hoping for a further continuation of the AIJ phase, and since AIJ

---

[60] GEF Office of Monitoring and Evaluation (2006). Climate Change Programme Study, GEF, Washington, also available at <http://www.gefweb.org/MonitoringandEvaluation/METhemesTopics/METClimateChange/2004_ClimateChange.pdf>.

[61] FCCC/SBSTA/2006/L/19/Add.1, 11 November 2006.

[62] <http://unfccc.int/resource/docs/2006/sbsta/eng/08.pdf>.

projects are being submitted for registration as CDM projects, this can be seen as a signal from the US of its willingness to keep the doors open for its return to the Kyoto Protocol.

The Clean Development Mechanism, launched in 1998 and operationalized in 2000, has registered 547 projects thus far. The largest number of projects are in India (32.54%), while the maximum number of credits accrue from projects in China. Some 51.91% of the projects are in Asia and the Pacific; 44.44 % in Latin America, while only 2.73% of the projects are located in the 53 countries of Africa, and 0.91% in other regions.[63] Investment in this mechanism tends to follow foreign direct investment trends and there are projects mostly in countries that are attractive to the host countries. In all, 48% of the projects focus on energy, but significant emission reductions are expected from the projects aimed at reducing HCFCs. While these projects are supposed to both support emission reduction and promote sustainable development, experience shows that since only the former is monitored and verified, in general only the directly related sustainable development benefits (i.e., those benefits that are directly related to emission reduction) are generally achieved and the indirect benefits are not.

The Climate Change Convention promotes technology transfer,[64] and the Subsidiary Body for Scientific and Technological Advice (SBSTA) coordinates efforts to do so. Traditional technology transfer runs the risk of transferring old, affordable and inefficient technologies to the developing countries, while technology transfer under the Convention aims to counteract this process by stimulating climate friendly technology transfer. Most of this transfer takes place via the financial mechanisms mentioned above. However, many countries are also trying to mainstream climate change in their development activities. Japan indicates that in 2004 the share of environmental expenditure in its ODA was 40% – up from 14% in 1994.[65] The US reported to the SBSTA that it actively engages in technology cooperation via the Asia Pacific Partnership on Clean Development and Climate, the Carbon Sequestration Leadership Forum, Generation IV International Forum, Global Nuclear Energy Partnership, the Group on Earth Observations, the International Partnership on the Hydrogen Economy and the Methane to Markets Partnership. This last Partnership is expected to save 180 million tonnes of CO_2 equivalent.

5.   CONCLUSIONS

This paper has argued that in terms of developing country issues, two elements stand out. The first is the possibility of China and India accepting some form of quantitative commitments under the evolving climate change negotiations; and the

---

[63] <http://cdm.unfccc.int/Statistics/Registration/RegisteredProjByRegionPieChart.html> (last visited on 15 March 2007).

[64] Art. 4.5 of the UNFCCC.

[65] FCCC/SBSTA/2007/2 at p. 5.

second is the impact of climate change and climate change politics on the larger group of developing countries.

There are strong indications that the US is keeping its options open.

In terms of the first issue, this article takes the position that it is most unlikely that China and India will accept quantitative targets, until and unless the US also does so. The key question is the following: is the US likely to return to the folds of the multilateral negotiating process? While a number of authors see this as unlikely, I would argue that there are five reasons why one can expect this to happen. First, the US is keeping its options open by pursuing a number of technology cooperation agreements with other countries. While some authors argue that such agreements only serve to undermine the multilateral process, especially as the US is at present challenging multilateralism in general,[66] the fact that the US states clearly in its State Department documents that it is undertaking measures consistent with its commitments under the Climate Convention,[67] and that it reports such activities to the SBSTA are indications that it is keeping a door open. Second, the US is investing in Activities Implemented Jointly (AIJ) and is promoting the extension of this programme's life. Since AIJ projects are being subsequently submitted to the CDM executive office for approval as a CDM project, the US may simply be trying to fast forward its participation in CDM as and when it becomes eligible to participate if and when it ratifies the Kyoto Protocol. Third, activities undertaken by states and cities within the US indicates a growing broad-based support for greenhouse gas reduction which more or less sends the message that if and when the US returns to the multilateral process, there will be some degree of nationwide support for it. Fourth, litigation in the US courts is adding pressure on the government. In April 2007, the Supreme Court by a vote of 5 to 4 decided in favor of the plaintiffs that the environment protection agency should regulate emissions from cars and this decision shows that the pressure is building up in the US.[68] Finally, as the Bush government becomes increasingly unpopular with Democrats and Republicans alike not only for its role in global security issues but also in relation to multilateral relations, it is not unthinkable that the next government will try to distance itself from current politics and, *inter alia*, return to negotiate a future regime. Already, there is internal pressure for reform. On 28 March 2007, the Senate Foreign Relations Committee passed a bi-party resolution from Democratic Senator Joe Bidden and Republican Senator Richard Lugar to return to the climate negotiations. Senator Biden reportedly said:

---

[66] Edward Newman, Ramesh Thakur and John Tirman (eds.), *Multilateralism Under Challenge? Power, Normative Structure and World Order*, UNU, Tokyo, pp. 289-307.

[67] FCCC/SBSTA/2007/2 at p. 6.

[68] Supreme Court of the United States, No. 05-1120, *Mass., et al.* v *Environment Protection Agency, et al.* on Write of Certiorari to the United States Court of Appeals for the District of Columbia Circuit, 2 April 2007.

'For too long we have abdicated the responsibility to reduce our own emissions, the largest single source of the problem we face today. We have the world's largest economy, with the highest per capita emissions. Rather than leading by example, we have retreated from meaningful, binding, multilateral international negotiations. With this resolution, we want to put the Senate on record in support of a new effort to build trust, to make commitments, and to participate in a coordinated international effort to confront the real threat of climate change.'[69]

Current pressure at the G-7 summit and possibly a proactive stance by China which has announced a climate policy in June 2007, possibly added to the momentum which led to the diplomatically formulated commitment that

'In setting a global goal for emissions reductions in the process we have agreed today involving all major emitters, we will consider seriously the decisions made by the European Union, Canada and Japan which includes at least a halving of global emissions by 2050.'[70]

Keeping up the pressure on China and India via mainstreaming in the UN, making cooperative agreements with local governments and through the potential threat of eventual domestic litigation is vital.

If this scenario is likely to occur, and keeping in mind that China and India themselves have a lot to lose from the potential impacts of climate change, these two countries need to ensure that they have not locked themselves into a technological and institutional trajectory which makes accepting commitments impossible. With China building a new power plant every week on average at present, this risk is great. It is here that one hopes that a consistent message from UN agencies and international banks will promote prudent commitment to sustainable development. Engaging the governments through the various climate change mechanisms is likely to keep the issue high on the national agenda. Engaging local governments to take action will help to create a broad base of support. All these should provide the necessary carrot. The threat of future litigation should perhaps act as the stick.

Keeping other developing countries engaged via the various funding, technology transfer and capacity building mechanisms is vital; and an increasing focus on adaptation is increasingly becoming an absolute necessity

The attention I have paid to China and India may seem disproportionate in a world of about 150 developing countries. That was not my intention. Engaging these two countries and making it possible for them to participate seriously is critically important for the rest of the developing countries. For without their participation, any

---

[69] Press Release, dated 28 March 2007, by Joseph R. Biden, Jr, <http://biden.senate.gov> (last visited 8 June 2007).

[70] G-8 Summit Statement on Climate Change, 7 June 2008, <http://www.foxnews.com/wires/2007Jun07/0,4670,G8SummitClimateText,00.html> (last visited 8 June 2007).

hope of addressing the climate change problem recedes hopelessly into the future. However, when we look at the case of the other developing countries, we see that the resources available to help them adapt are pitiably small, mere token efforts; especially in comparison to the sort of resources they claim is necessary for adaptation measures to serious mitigation measures. Should the CDM turn out to be highly successful, there may be more resources in the future. It is ironic, however, that help for vulnerable countries comes from a tax on North-South cooperation. In terms of understanding whether any of the climate change mechanisms contribute to the development of the poorer 100 countries, one can only conclude that very few resources are channeled into 52 African countries (excluding South Africa) or to the small island states.

While mainstreaming climate change in the UN will have the effect of sending an internally consistent message to these other developing countries, it is also important that any long-term regime on climate change is based on a predictable system of commitments, and not merely on the whims and interests of individual countries. Such a system should not be too complicated – like the emissions trading system – but understandable and implementable by all countries.

As global dynamics change, the axis of power is changing – but the poor remain the neglected and most vulnerable; vulnerable to the lack of action in the developed and the rapidly developing world, but often also to their own incompetent and unaccountable governments.

CLIMATE CHANGE, SECURITY AND FORESTS

Wouter Veening*

Climate change is increasingly seen also as a security issue, as evidenced by the discussion on this topic in the Security Council of 17 April 2007 after a proposal by the UK.[1] Floods and droughts, rising sea levels, the loss of arable land, the spreading of diseases over larger areas, amongst other things, will result in competition for scarce resources, mass migrations, food and water insecurity and thus the potential for violent conflict within and between states. The conflict in the Middle East is to a large extent a water conflict, but has not yet been recognized as such. If the glaciers on the Tibetan Plateau keep melting the water supply of Pakistan, India, Bangladesh, the Mekong countries and China will be seriously compromised. Most of these countries are densely populated and some of them are nuclear powers.

While, of course, everything has to be done to mitigate further climate change, adaptation to inevitable change is of the utmost importance. Forests, especially tropical forests and mangroves, play a crucial role in both mitigation and adaptation. They store carbon and sequester CO_2, they break the speed of hurricanes and typhoons, they prevent mud-slides, and mangroves protect against rising sea levels. Conversely, deforestation and especially forest fires cause 25% of greenhouse gas emissions.

The day after the conference at the Asser Institute that resulted in this publication, the author traveled to Kalimantan, the Indonesian part of Borneo, to discuss the progress of the Central Kalimantan Peatland Project (CKPP).[2] CKPP has as its main objective to prevent further drainage, deforestation and fires in the peatlands of the Province of Central Kalimantan, which together with the deforestation, drainage and fires in other parts of Indonesia, make that non-industrialized country the third emitter of greenhouse gases globally, after the USA and China. The reason for this is that peatlands store huge quantities of organic material, equivalent to approximately 2,000,000 tonnes of CO_2, comparable with 100 years of the current emission of fossil fuels. The fires also cause serious smog in the surrounding countries, with great damage to public health, the impossibility of air traffic and problems for navigation in the Straits of Malacca

The causes are slash-and-burn methods employed by local farmers, illegal logging and the establishment of oil palm plantations on illegal sites, such as lands

---

* Director of the Institute for Environmental Security, The Hague.

[1] <http://www.un.org/News/Press/docs/2007/sc9000.doc.htm>.

[2] The project's website is <http://www.ckpp.org>.

W.Th. Douma et al., eds., The Kyoto Protocol and Beyond
© 2007, T·M·C·ASSER PRESS, *The Hague, The Netherlands and the Authors*

with a peat layer of more than 3 meters. The main driving forces are the demand for timber and palm oil from China and the European Union, not only for the food industry but now also as a replacement for fossil fuels. In the European Union, Directive 2001/77 even prescribes that in 2010 at least 5.75% of all fuels should come from biomass.[3] One tonne of palm oil produced on peatlands, however, causes an emission of 20 tonnes of CO_2, so from a climate perspective the balance is very negative.[4]

Commissioned by the World Wide Fund for Nature (WWF) a suitability map for oil palm plantations in Borneo has been made by SarVision, an advanced monitoring consultancy which came out of Wageningen University, and it concludes that the production of palm oil should be limited to those areas indicated as suitable on that map. A strict and independent monitoring regime, as already developed by SarVision, should of course be applied.

From neighboring countries pressure is now being exerted on the Indonesian President and his cabinet to prevent another season of fires. An operational plan, possibly involving the army, should be worked out together with Indonesia's neighbors and also with a country such as the Netherlands which is willing to assist technically and financially, but only if there is a viable plan. As Indonesia will host the 13th Conference of the Parties of the UNFCCC in December in Bali, it is thought that it would be very awkward for the Indonesian government and authorities if the meeting would take place during or just after a severe fire.

With the authorities and local communities in Indonesia and elsewhere in the world contracts should be concluded to compensate them for their efforts in maintaining the functions of ecosystems such as forests in the field of climate stabilization, the preservation of biodiversity and the protection of water flows, the so-called payments for ecosystem services. These contracts would break away from the principle of 'common but differentiated responsibilities' and instead would be a form of equal partnership, more befitting in a world where everybody has the responsibility to do whatever one can to stop further climate change and assist each other in adapting to unavoidable change.

---

[3] Directive 2001/77/EC of the European Parliament and of the Council of 27 September 2001 on the promotion of electricity produced from renewable energy sources in the internal electricity market, OJ 2001, L283/33. In its Communication 'An Energy Strategy for Europe' (COM(2007) 1 final) of 10 January 2007, the Commission introduced the plan to set an even higher binding target of 10% for the share of biofuels in petrol and diesel in each member state in 2020. This time, however, the target is to be accompanied by the introduction of a sustainability scheme for biofuels.

[4] In the Netherlands, attention to the unsustainable nature of some forms of biomass resulted in the development of a set of framework criteria for sustainable biomass production ('Toetsingskader voor duurzame biomassa'. Eindrapport van de projectgroep 'Duurzame productie van biomassa'). In the letter dated 21 December 2006 offering this report to Commissioner Dimas of DG Environment, it is underlined that developing sustainability criteria for biomass production and use in individual member states of the EU is not the most effective approach, and that the development of such criteria at the EU level would have an important added value. *Source*: Dossier Biobrandstoffen at <http://www.minvrom.nl>.

A regime to protect forests cannot wait until 2012. Action to prevent irreversible damage to the global climate and loss of biodiversity should be taken now. But it should, of course, become part of the post-2012 arrangements. Also, countries such as China and India should be Parties with fully-fledged responsibilities. Last but not least, the USA and Australia should of course join.

EU ETS IN THE POST-2012 REGIME: LESSONS LEARNED

Chris Dekkers* and Machtelt Oudenes**

1. INTRODUCTION

Under the Kyoto Protocol, the European Community committed itself to reducing its greenhouse gas emissions by 8% during the period 2008-2012 compared to the emission levels of 1990.[1] In its Article 17 the Kyoto Protocol introduced three flexible mechanisms to facilitate cost-effective emission reductions.[2] A primary role was defined for Emission Trading (ETS) as the most important instrument to achieve the emission targets for the Annex I countries. Parallel to emissions trading, two other instruments were agreed upon and introduced, i.e., Joint Implementation (JI) and Clean Development Mechanisms (CDM). The very idea behind emissions trading is that it supports and actually forces industry to seek cost-effective emission reductions in their own installations and to develop new, less polluting production processes. If properly set up and designed, emissions trading is thought to be much more capable of realizing the very ambitious reduction targets required from the industrial and other sectors towards the kind of changes needed for future Climate Change policies.

As soon as the Kyoto Protocol had been agreed among the EU member states, in 1999 the European Commission already took the first steps towards developing emissions trading. Working Groups were set up to discuss and design the various elements of the emission trading scheme. Following the publication of the draft directive in 2002 and an astonishing fast-track negotiation process, full agreement on the Directive on CO_2 emissions trading[3] was reached in the Council of Ministers and the European Parliament in October 2003. At that moment in time few people if any fully realized what emissions trading would entail and how much

---

* Ministry of Housing, Spatial Planning and the Environment (VROM), the Netherlands.
** Meurs Juristen Consultant, the Netherlands.
The authors would like to emphasize that the views and opinions presented in this article are their own, and do not necessarily represent the views or opinions of the Ministry of VROM.

[1] Green Paper on greenhouse gas emissions trading within the European Union, Brussels 8 March 2000, COM (2000) 87 final.

[2] Kyoto Protocol to the United Nations Framework Convention on Climate Change, United Nations, 1998 at: <http://unfccc.int/kyoto_protocol/items/2830.php>.

[3] Directive 2003/87/EC of the European Parliament and the Council of 13 October 2003 establishing a scheme for greenhouse gas emission allowance trading within the Community and amending Council Directive 96/61/EC.

more discussions would be needed to obtain sufficient clarity on a large number of legal and regulatory issues. Although internal discussions in the Commission Services and in small working groups of government and industry experts already started in 1999, and in 2001 a much wider range of member states and industry experts had been involved in the development of the draft directive, these early discussions were primarily focused on the more economic elements of an emission trading scheme, such as the principles of allocating emission allowances and the issues of how to set a national cap and to accommodate economic growth. Member states, industry and other stakeholders were much less preoccupied with the more practical and judicial issues of the scheme. Moreover, some of the more intricate elements of the judicial and compliance framework of emissions trading were overlooked, disregarded or marginalized at that time.

Another element, barely recognized at that moment, was that changing from a command and control regulation to a market-based instrument requires a totally different way of thinking. Emission trading is an all-embracing instrument with quite different linkages to private law, international law, administrative law and penal law. The complexity of this new market-based instrument and the range of legal, technical and practical issues that had to be clarified at the start of the scheme, initially led to varied reactions by the member states, sectors of industry and other stakeholders. Most of these issues and problems have been solved along the way during the implementation of the emissions trading scheme, while other issues still need clarification or resolving. Moreover, to make the 'grand experiment'[4] work as intended and to make it deliver the emissions reductions agreed under the Kyoto Protocol and, even more so, to achieve the ambitious emissions reductions needed for the period after 2012,[5] the EU Emissions Trading Scheme (EU ETS) needs to be evaluated as to its effectiveness, reviewed as to its functioning and adjusted accordingly.

This article will focus on the lessons learned for the post-2012 period and address the possible implications and adjustments needed for the future legal and regulatory framework to achieve these ambitious 20-30% greenhouse gas emission reductions agreed in the European Council of Environment Ministers.[6] The central question is what lessons can be drawn from the first trading period and to what extent these lessons are still relevant for a proper functioning of emissions trading in the future. Furthermore, what adjustments are needed with respect to allocation, compliance and legislation? This paper will explore to what extent these issues are covered in the present discussions on the review of the EU ETS directive in view of the post-2012 period.

---

[4] J.A. Kruger and W.A. Pizer, 'Greenhouse Gas Trading in Europe, The New Grand Policy Experiment', *Environment*, Vol. 46, No. 8, October 2004.

[5] Communication from the Commission to the Council, the European Parliament, the European Economic and Social Committee and the Committee of the Regions, Limiting Global Climate Change to 2 degrees Celsius. The way ahead for 2020 and beyond, Brussels 10 January 2007, COM (2007) 2 final.

[6] Presidency conclusions of the European Council, Brussels, 8-9 March 2007 (conclusion 32).

2. THE CRITICAL ISSUES OF EMISSIONS TRADING

The introduction of emissions trading implies a drastic departure from the command and control regulatory framework that has been at the centre of traditional environmental legislation in Europe. In a market-based environment, such as emissions trading, the role of the competent authority becomes quite different compared to that in a traditional command and control regulatory environment. In a command and control 'environment' competent authorities lay down emission limit values in permits or in general binding rules and enforce these emission limit values through inspection and often rather lenient enforcement actions. Only in exceptional situations when companies continue to fail to meet those limits, will the competent authorities take drastic enforcement steps and apply severe sanctions.[7] In an emissions trading scheme, however, competent authorities operate in a different regulatory environment, with an overall emission target 'distributed' over all regulated subjects and with very precise, and uniformly defined rules on permitting, monitoring and reporting on the basis of which emissions trading is to function. Subsequently, it is basically left to the companies themselves to find the means to achieve those individualized emission targets. As a result thereof the role of the competent authority shifts towards checking whether the companies have indeed monitored their emissions correctly and accurately in accordance with the detailed rules and regulations, whether the emission data have been verified and whether emission allowances equivalent to the reported emission data have been surrendered.

Another aspect, overlooked or not fully recognized at the time when emissions trading was introduced, is that EU ETS is not just a cost-effective environmental instrument, but is seen and treated by the industrial and financial sectors as primarily a financial instrument. This means that apart from the environmental objective[8] as defined by the government, emissions trading in its practical functioning is also determined by commercial and financial objectives. These financial and economic objectives have a direct bearing on the way emissions trading should be designed and regulated as well as on the safeguards that need to be inserted in the system from the very start. In fact the proper functioning of the emissions trading market very much depends on the timely and accurate information provided to the market participants, and on the rigour with which the monitoring and reporting of emissions has been set up, as well as the strictness with which the competent authorities enforce compliance.[9] Moreover, new concepts have been introduced, which were not earlier applied in environmental law: transparency, accountability and integrity have become crucially important, as have the elements of trust and confidence in the market.[10] The major challenge for the third, post-2012 trading period

---

[7] J.A. Kruger and C. Egenhofer, 'Confidence through compliance in Emission Trading Markets', prepared for the International Network for Environmental Compliance and Enforcement (INECE) 17-18 November 2005, p. 3: at <www.inece.org>.

[8] Consideration 3 in Directive 2003/87/EC.

[9] J. Kruijd, et al., 'Building Trust in Emission Reporting, Global Trends in Emission Trading Schemes', PricewaterhouseCoopers, February 2007.

[10] J.A. Kruger and C. Egenhofer, see *supra* n. 7, p. 4.

will be how to ensure and enforce the transparency of the system, the accountability of the market participants and market integrity. In order to achieve that, the EU ETS will need a thorough review of some of the legal provisions, an evaluation of the responsibilities and tasks of the parties, and a careful examination of the regulatory requirements of inspection and enforcement by the member states.

3. FACTORS THAT DETERMINE THE SUCCESS OF EU ETS

The success of EU ETS is to a large extent determined by clear structures as well as a strict definition and a delineation of the roles and responsibilities between different parties in EU ETS. Not only a clear delineation of the roles and responsibilities of the European Commission and the member states is necessary, but also of the competent authority and the operator so that it is absolutely clear who is responsible for what and who can be held accountable. The same is true for the verifier and the accreditation bodies accrediting those verifiers. One of the crucial elements of emissions trading and subsequently determining its success or failure is a clear-cut and well-defined division of tasks and responsibilities in the public and in the private domain. It is vital to ensure a clear distinction between the two domains in order to avoid too much interference from the public domain authorities in the market as this will undoubtedly affect the trust that market participants will have in the market process. A verifier should not redo the task of the competent authority by checking whether the monitoring plan was in line with the rules and regulations of the Monitoring and Reporting Guidelines.[11] In a similar way the competent authority should not redo the verification of emission reports, as that was defined as being the task of the verifier. However, it is also of critical importance that proper structures, procedures and standards in the entire EU are in force so as to ensure that in the private domain the various tasks, functions and responsibilities are well executed and accounted for, and that the market participants adhere to the principles of transparency, accountability and integrity.

A second, most important aspect for a proper functioning of the market of emissions trading is the way emission allowances are allocated. In the first trading period the National Allocation Plans[12] of the member states differed substantially from one another in content, principles and actual allocations of emission allowances to individual sectors of industry, thereby creating major diversity in the allocation processes.[13] Furthermore, the total emission allowances to be allocated to industrial installations were not set in a sufficiently stringent and constrained framework by the Commission. At the same time most member states were concerned not

---

[11] Commission Decision of 29 January 2004 establishing guidelines for the monitoring and reporting of greenhouse gas emissions pursuant to Directive 2003/87/EC of the European Parliament and of the Council, 2004/156/EC, OJ L 59/1.

[12] Art. 9 of Directive 2003/87/EC.

[13] Application of Emission Trading Directive by EU member states, European Environment Agency (EEA) Technical Report, No. 4/2007.

to disadvantage their own national industries. The Commission has provided extensive guidance on the drafting of National Allocation Plans (NAPs) for the second
trading period, and this in itself was a clear improvement towards more clarity and
uniformity in the allocation process. However, the second period NAPs in general
still differ widely in outline, transparency and in certain allocation principles. Moreover, a number of allocation issues remain unresolved.

Thirdly, strict compliance with and rigorous enforcement of the monitoring, reporting and verification requirements of the scheme are crucial to a proper functioning of emissions trading. Accurate and correct emission data require strict and
uniform monitoring and reporting requirements, not only uniformly applied to the
industrial installations in one country, but in the end also uniformly and equally
applied to all industrial installations in all the member states. There is a long way to
go to attain more harmonization of these requirements. So far, information on the
quality of emission data from most member states is insufficient and lacking, and
unless major progress is achieved towards much greater transparency concerning
those data, emissions trading will not assure the trust and confidence that the market needs. Moreover, the first trading period has clearly demonstrated the need for
enhanced inspection of the compliance of industry with the monitoring and reporting requirements, as well as the need for competent authorities to strengthen their
enforcement activities. In addition, proper communication between the competent
authorities and industry is crucial to ensure that industry understands the importance of compliance and enforcement actions within emissions trading and that
there is a higher level of acceptance for stringent compliance among the different
industrial sectors. Early and continuous communication between the competent
authorities and industry on inspection and enforcement strategy is therefore needed
to reach that level of acceptance.

4.   LESSONS LEARNED FROM THE FIRST TRADING PERIOD

During the early policy discussions on the Directive in 2002 and 2003, and again
during the following implementation process of EU ETS in the national legisla-
tions, most member states paid very little attention to the crucial importance of a
well developed and implemented system of monitoring, reporting, verification, inspection and enforcement. At that stage these elements of the compliance chain
were generally underestimated and undervalued by practically all parties. First of
all, this was due to the fact that in Europe in general there was and still is very little
experience with strict compliance in environmental issues. Secondly, the political
aspects of emission trading overshadowed more or less all other elements during
the preparations of and the discussions on the EU ETS directive. And, thirdly, the
discussions on the political, legislative and environmental objectives of the scheme
were thought to be much more relevant than the 'technical' details of monitoring,
reporting and enforcement. Member states and the European Commission had to
learn the hard way. In fact when the first discussions on the monitoring and report-

ing guidelines started in May 2003, the Commission and the member states were not fully prepared or sufficiently committed to participate in the discussions on the various technical details of CO_2 emissions monitoring and reporting, and had to gain the necessary experience. During the first trading period member states gradually learned the importance and financial relevance of the EU ETS directive, and along the implementation process they became aware of the need for much better cooperation and harmonization between the Commission and the member states. A number of lessons can therefore be drawn with respect to the various elements of the compliance chain.

4.1 Monitoring and reporting

From the very start the subsidiarity principle played a major role in the political discussions on the EU ETS directive.[14] As a result thereof the directive allowed member states a significant degree of discretion in its implementation. Moreover, certain requirements of the directive and the Monitoring and Reporting Guidelines initially led to quite different interpretations by the member states.[15] Moreover, both legal instruments allowed member states also to rely to a large extent on their existing national permitting and compliance structures and procedures. Hence most member states started the implementation phase of the directive with the intention to adhere as much as possible to their existing IPPC[16] permitting and compliance structures, procedures and standards. However, during the various implementation stages most member states gradually discovered that these 'command and control' structures and procedures did not fit well with the requirements and objectives of the EU ETS directive, and that they had to make a number of more or less drastic adjustments. Some member states, such as the Netherlands, Finland and others, decided early on to set up new structures for enforcing the permitting and monitoring requirements and to develop new procedures and standards to accommodate the EU ETS directive. Other member states, e.g., Germany, chose to make less drastic changes and adhered as much as possible to their existing IPPC compliance structures and procedures. Finally, the lack of experience with strict compliance has led to much confusion on the interpretation of the Monitoring and Reporting Guidelines (MRG), enacted in the Commission Decision of 29 January 2004.[17] At that

---

[14] During the negotiations on the EU ETS directive regular reference was made to Art. 5 of the EC Treaty establishing the European Community, which says that '*the Community shall take action, in accordance with the principle of subsidiarity, only if and so far as the objectives of the proposed action cannot be sufficiently achieved by the Member States and .....be better achieved by the Community*'. Although during these discussions it was recognized that emissions trading must follow the provisions and regulations of the internal market, member states tried to keep as much control as possible on the allocation of emission allowances and the development of the NAPs.

[15] J.A. Kruger and C. Egenhofer, see *supra* n. 7, p. 9.

[16] Council Directive 96/61/EC of 24 September 1996 concerning integrated pollution prevention and control, OJ L 257/26, 10.10.1996.

[17] Commission Decision of 29 January 2004 establishing guidelines for the monitoring and reporting of greenhouse gas emissions pursuant to Directive 2003/87/EC of the European Parliament and of the Council, 204/159/EC.

time it was not completely clear to most experts what the legal 'impact' and the structural and technical implications of the MRG were. Also some of the MRG provisions were rather unclear and ambiguous. As a result thereof the Commission Decision was implemented quite differently in the member states while ambiguities caused further discrepancies and diversity in the way monitoring and reporting have been actually carried out in the member states.

The diversity of procedures and requirements on the permitting process in the first trading period was characterized also in different degrees of stringency with which the monitoring plans of installations were validated throughout the EU. Some member states validated monitoring plans before issuing a permit; others validated monitoring provisions after issuing the permit or performed only marginal checks. Also some of the definitions laid down in the directive led to major confusion. Especially the reference to '*installation as a stationary technical unit where one of the activities listed in Annex I of the directive is carried out*' was cause for misunderstanding among member states. As a result thereof the scope of the EU ETS directive was interpreted differently throughout the EU. Some member states applied, for example, a broad definition of combustion installation, whereas other member states applied a more narrowly defined concept of combustion installation, thereby limiting the scope of the directive. This also complicated the monitoring of the entire industrial installation since in a number of situations the narrow interpretation of the combustion plant will cause only part of the installation to fall under the emission trading scheme. This influences also the accuracy by which emissions covered by the scheme are monitored. The question of applying a narrow or broader definition of an installation is an issue that needs to be resolved well before entering the third trading period as it impacts the scope of the monitoring of combustion emissions as well as the effectiveness of the whole emission trading scheme.[18]

Other discrepancies in interpretation related to the technical (in)feasibility and the definition of unreasonable costs. Practical experience in the coming years will need to clarify in which situations an operator is or can be allowed to deviate from the monitoring requirements because of unreasonably high costs and/or implied technical infeasibility to meet the requirements.

It has become clear to most stakeholders that to overcome these different interpretations of the directive and the monitoring and reporting guidelines, much more harmonized structures, procedures and standards as well as more uniform monitoring and reporting requirements are needed. The major question to be resolved for the next trading period is how to ensure sufficient harmonization and how to guarantee that in all member states monitoring plans are validated by the competent authorities in a similar consistent manner. One solution would be to formulate the

---

[18] During the first trading period the Commission rejected the narrow interpretation of the term industrial installations in the NAPs proposed by France and Spain. During the recent review stakeholders discussion under the second European Climate Change Programme (ECCP-II) representatives of member states proposed to use a common, broad definition of installation for the third trading period, so that all combustion plants and CO_2 process units within the industrial site would fall under EU ETS.

current Commission Decision on monitoring and reporting into a regulation ensuring a direct legally binding effect for the subjects it aims to regulate.[19]

4.2 Verification and accreditation

A similar type of confusion occurred with respect to the requirements for verification and accreditation and the way appropriate structures and procedures had to be defined. These two elements of the compliance chain did not receive much attention during the early technical and political discussions on the EU ETS directive. The implementation of the directive in 2005, the verification of the first emission reports in April 2006 and the publication of the emission data during that month in the member states showed a similar diversity in procedures.

Because of the rather vague and broad phrasing of the verification requirements in the Monitoring and Reporting Guidelines and Annex V to the EU ETS directive, different verification structures, procedures and standards have been set up in the member states.[20] The guidance of the European Cooperation for Accreditation, i.e., the EA Guidance for Recognition of Verification Bodies under EU ETS directive, which contains more detailed requirements on the verification process and the competence of the verifier, is not a legally binding document and thus not sufficient to enforce uniformity in the EU.[21] As a consequence thereof, a number of member states developed their own verification and accreditation procedures and requirements resulting in a situation that there are no harmonized verification and accreditation structures or standardized requirements for accreditation bodies in the EU. Consequently no real control mechanisms exist that can check the uniform application of verification requirements and practices in the member states.

For the market to function properly and to ensure the public's confidence it will become increasingly important that the public domain, i.e., the Commission, governments, competent authorities, on the one hand, and the private domain in which operators, verifiers and accreditation bodies operate, on the other, are well defined, well structured and operate on the basis of distinct and properly defined requirements. Such a strict division of tasks and responsibilities between the public and the private domain is only possible if the manner in which verification and accreditation is carried out is properly safeguarded with strong control mechanisms so that the quality of the verifier can be checked. Without these checks and uniformity of requirements, verification will not provide EU ETS with the safeguards that are needed.

---

[19] Several representatives of the EU member states have suggested in the review stakeholder discussion meeting under ECCP-II to regulate the provisions laid down in the Monitoring and Reporting Guidelines in a Regulation in order to ensure a harmonized implementation of these requirements.

[20] H. Schoolderman and M. Carrington, 'Verification: Assuring the Credibility of the European Union Emissions Trading Scheme', PricewaterhouseCoopers, 2005.

[21] EA-6/03, EA Guidance for Recognition of Verification Bodies under EU ETS Directive, European Cooperation for Accreditation, March 2007.

As said before, how these safeguards of verification were to be achieved was not really addressed nor foreseen during the drafting and the decision-making process of the EU ETS directive, as a result of which a patchwork of different mechanisms and requirements was set up in the member states. In view of the financial risks and interests involved, robust and harmonized structures and procedures will be needed to attain sufficiently strict compliance. Such robust verification and accreditation schemes can be achieved by authorizing the Commission to submit and define verification and accreditation requirements in a regulation. A harmonized application of a EU accreditation framework should be set up to ensure that foreign verifiers will be allowed to verify emission reports in other member states without having to be accredited once again by national accreditation bodies. Finally, to realize a much stronger and more effective accreditation role it should be considered to what extent a centralized body for the quality assurance of accreditation bodies should be set up. This would enhance the quality of verification and ensure uniform application of the verification process. In that context major questions need to be addressed such as: how do we ensure that the controls in the private sector are functioning well, what structures should be set up to guarantee a proper outcome, what additional standards are needed, which procedures and requirements need to be defined, and, finally, to what extent should the private sector be encouraged or challenged to ensure that its own house is better organized? It should also be assessed which tasks and responsibilities should remain in the public domain, and also whether the private domain could really function without some kind of public oversight.

4.3   Inspection and enforcement

Other lessons learned relate to the way member states have used inspections and sanctions to enforce compliance. Right now there is no uniformity in the way sanctions are applied by member states, as well as the way and the extent to which compliance and enforcement are being carried out. Moreover, it is to be expected that when moving from the second trading period towards the third period the price of a CO_2 allowance will rise considerably due to the ambitious emission reduction target which the Council of Ministers agreed for the period after 2012. This increase in the price for CO_2 allowances will among other things intensify the pressure from industry on national governments and competent authorities which will increase the risk of cheating and fraud.

Strong and strict compliance mechanisms as well as uniform requirements should therefore be the 'rule' and not the exception. According to the US Environment Protection Agency (EPA) handbook on enforcement 'compliance strategy' must ensure that 'requirements are met and the desired behavior is achieved'.[22] The first condition is that credible enforcement is developed through a set of government actions to compel or encourage compliance. The second condition is that real deterrence is created, i.e., an atmosphere where industry and other actors choose to com-

---

[22] Principles of Environmental Enforcement, US EPA, 1992, EPA-300-F-93-001.

ply rather than to violate the law. The third condition is that the deterrence is effective, which implies that there is a credible likelihood that any violation will be detected, and that upon detection the competent authority will act swiftly and respond to the violation by imposing sanctions or penalties. Finally, among all the 'regulated subjects' the perception must exist that the first three conditions apply, i.e., credible enforcement, high probability of detection and certain deterrence by sanctions and penalties.

The uniformity and harmonization that is required to meet these three conditions and thus to achieve a strict compliance and rigorous enforcement system is a new phenomenon in European law. Inspection and enforcement have long been regarded as the 'natural domain' of the member states and the most prominent example of subsidiarity. Given the financial, environmental and commercial interests involved and the importance of an adequate enforcement to ensure strict adherence of the regulated subjects to the requirements of the system, 'compliance' in emissions trading is however quickly becoming an issue of a level playing field, of internal market considerations and therefore an issue of harmonization. Major steps have still to be made towards achieving acceptance by industry and towards uniformity of enforcement in Europe. This is an issue of great relevance in the review process of the EU ETS directive.

4.4 Sanctions

More harmonization would also be necessary with regard to the application of sanctions in emissions trading. For example, a list of effective sanctions per infringement in the directive would be helpful in attaining a level playing field. Furthermore, the new sanctioning mechanisms like naming and shaming require special attention. Naming and shaming is quite common in the United States and in the financial markets sector. According to Article 16, paragraph 2 of Directive 2003/87, member states shall ensure the publication of the names of operators who have breached the requirements to surrender sufficient allowances under Article 12, paragraph 3 of Directive 2003/87. However, this could raise legal problems in the EU when the presumption of innocence principle laid down in Article 6 of the European Convention on Human Rights (ECHR) has not been met. If the competent authority in its public opinion determines directly and without reservations that the operator is in breach of the legal requirement to surrender sufficient allowances, and moreover without having properly checked that the operator is indeed not in compliance with the requirements, the presumption of innocence principle will be violated and Article 6 ECHR will not be complied with. If challenged in court this issue could affect the effectiveness of the naming and shaming provisions in the directive.

4.5 Exchange of best practice

All the compliance elements, i.e., permitting, monitoring, reporting and verification up to inspection, enforcement and sanctions are interconnected and equally important to ensure the proper functioning of the whole compliance chain. One of the first steps and crucial elements towards creating a higher level of harmonization is exchanging information and best practice between different countries. Likewise, this element was not sufficiently recognized at the start of the EU ETS. For instance, a range of lessons can be learned from the robust, efficient and effective compliance and enforcement mechanisms that the EPA in the United States has developed since 1994 in their emissions trading programmes.[23] Moreover, in the latest versions of their control mechanisms extensive use is being made of new information technologies instruments. Other lessons can be drawn from best practices in other member states, as well as from the financial world. It would be advisable to use the experience of the financial sector in developing the control environment that is common practice among financial institutions and to learn how they have developed new information technology to achieve a high performance in financial compliance.

As major financial interests and risks are involved in emissions trading strict rules are, for example, necessary for the publication of emission data. Publication of incorrect data or improper procedures to make data publicly available can have a major financial impact. For instance, in April 2006 the price of CO_2 allowances fell in a few days from € 30 to € 15 and to much lower levels months later. This was mainly due to the limited knowledge of market participants concerning the number of allowances issued in relation to the actual emissions, but partly also because competent authorities in the member states had no experience with the sensitivity of publishing emission data in relation to financial data and the implications for a financial market. In addition, the transparency of emission data required under the Århus Convention may cause tensions with the implications that the disclosure of information may have on the financial market.

5. OTHER LEGAL AND TECHNICAL ISSUES OF EMISSION TRADING

Of course, the success of emissions trading is not solely and only determined by the effectiveness of the compliance chain. Other legal and technical issues and questions are closely connected to the proper functioning of the market.

5.1 Allocation

As is well documented, to a large extent the efficiency of emissions trading in the first trading period has been hampered by the diversity in the way member states have allocated emission allowances to the national industry and how they have set up their national allocation plans. To some extent this is also true for the allocation of emission allowances in the second trading period, even though the Commission supervised the allocation more strictly, carefully evaluated each national allocation

---

[23] J.A.Kruger and C. Egenhofer, see *supra* n. 7, p. 15.

plan and in nearly all cases made adjustments towards lower allocation levels. Apart from the overall number of allowances to be allocated to their national industry, member states struggled also with the principles of how to allocate, in an appropriate manner, allowances to new entrants, how to incorporate economic growth rates into the national allocation plans and how to deal with closures. Should allowances be withdrawn in the case of a closure, on what legal basis and who should get the allowances from a closed plant?

Another issue was the low price of CO_2 allowances as a result of over-allocation by most of the member states, and the impact this over-allocation had on the proper functioning of the market. During the last couple of months the price of CO_2 for the first trading period has fallen to € 0.50 per tonne. Because of a much more stringent set of allocation principles defined by the Commission, and a much tougher line in approving the NAPs of the member states, the price for the second period allowances is steady at a more realistic level of € 15 per tonne. As very ambitious reduction targets have been formulated for the period beyond 2012, a major question becomes how to allocate allowances to industry ensuring that by 2020 an overall 20-30% emission reduction is achieved. Some member states prefer benchmarking as a means to allocate while others are more inclined to auctioning emission allowances. Closely connected to this question is whether national allocation plans should still be maintained in the third trading period along with strict and effective guidance from the Commission. There is a real danger that if member states must develop their NAPs in a similar procedure as used for the second trading period, they will have great difficulty in withstanding the pressure from their national industry. By now there is a clear understanding that a new, better allocation methodology should be used.

At the moment of writing this article benchmarking and auctioning are at the top of the list in the review process. How justified and economically attractive benchmarking may be at first sight, the problem with benchmarking is the difficulty in agreeing on the technical assumptions used as departing points for the discussion, and the different implications benchmarking will have for individual companies. It may prove to be extremely difficult to reach a political agreement between the new and the old member states since there are likely to be major differences in the technology applied with probably unwelcome consequences for allocation if benchmarking would be used. In the discussions on the review of the Directive the interest is growing for auctioning as a means to avoid favouring national industries and windfall profits. At this moment European industry is in general not keen on auctioning. Therefore it will not be easy to get agreement on the use of this allocation methodology.

However, the inadequateness of allocation on historical emissions has been demonstrated by the windfall profits in the power sector. The discussion in Europe on these windfall profits has led the RGGI states in the US[24] to choose for 100%

---

[24] The Regional Greenhouse Gas Initiative, or RGGI, is a cooperative effort by the North-eastern and Mid-Atlantic states of the US to reduce carbon dioxide emissions. The RGGI participating states will be developing a regional strategy for controlling GHG emissions. Central to this initiative is the implementation of a multi-stage cap-and-trade emissions trading scheme for the power industry.

auctioning in their CO_2 emissions trading programme. Despite industry's reluctance in Europe, auctioning is most likely to play an important role either combined with benchmarking or with allocation based on historical emissions. Linking an EU-wide cap with partly benchmarking and partly auctioning could be a possible way out to make the allocation more transparent and function better. However, auctioning also immediately raises the issue of who is to 'get the proceeds', and for what purpose the income from auctioning should be used. It is obvious that the allocation methodology to be used for the third period will become a highly politicised issue, and will involve extensive political discussions before being resolved.

5.2   Administrative burden of installations

Another lesson concerns the administrative burdens of the scheme. The costs implied by the requirements for the monitoring, reporting and verification of emissions are generally considered to be too high, especially for small installations emitting less than 25,000 tonnes of CO_2. On a number of occasions member states raised this issue in the preparations for the second trading period NAPs and tried to convince the Commission to formulate legal or procedural proposals that would exclude small installations from the scope of the directive or introduce a threshold to exempt them from the monitoring requirements. Several measures have been taken to reduce the administrative burden for small installations by simplifying the monitoring requirements. However, for the third trading period it should be assessed whether these measures are sufficient. Moreover, additional evaluations and the assessment of cost-reduction options will most likely be necessary on the monitoring efforts for major, minor and very small source streams.

5.3   Legal status of an emission allowance

Another feature of emissions trading is that it involves the introduction of new legal concepts. During the implementation of the directive in national law the legal status of an emission allowance became an issue: should an emission allowance be regarded as a property right or not? How does one deal with these new property rights became one of the major legal questions in the implementation process of the EU ETS scheme in national law. The answers to these questions were and still are highly relevant for determining under what conditions emission allowances can be transferred between private parties. Most of these issues have been sufficiently addressed and answered. However, there are still issues that need to be resolved.

5.4   Interrelationship with IPPC directive

One of the issues was also how emission trading would relate to the IPPC directive and which amendments would be needed to enable an effective and full implementation of emission trading. Nearly all installations falling under the emission trading scheme are also covered by the IPPC directive. This means that installations that are subject to both directives must have an IPPC permit as well as an emission permit. According to Article 26 of Directive 2003/87/EC concentration values which relate to direct CO_2 emissions shall not be included in the IPPC permit unless this is necessary in order to prevent significant local environmental problems.[25] By submitting this exception in the directive the full implementation of emission trading is not harmed by concentration values and emission limit values in the IPPC directive. Again no full agreement and common understanding on the practical implications have yet been reached. It is very likely that in the future this issue will have to be clarified in court decisions.

5.5   Transparency of monitored data

Other legal issues arose with regard to the transparency of monitored data. Although emission data are public according to the Århus Convention on the public access to environmental information,[26] some underlying emission data may be considered as confidential information. This raises the issue of how to deal with the right of the public to information versus the right of individual companies to keep certain information confidential. The future will probably teach us where to draw the line.

5.6   Lessons learned on consultation and cooperation

All the aforementioned questions had to be addressed and answered in some harmonized way. In the past member states addressed the issue of how to interpret directives within their 'national context'. They would find solutions that they felt most appropriate in their own national situation. As a matter of fact, the types of questions and implementation problems described in this article had not really come to the surface in earlier 'implementation' processes with other directives since the diversity in implementation between member states did not cause major differences in the competitive position of national industries and were not considered that relevant or crucial in the more traditional command and control regulatory schemes. The implementation of the EU ETS directive, however, necessitated much more thorough, in-depth political and technical discussions between the member states and the Commission than for any other directive in the field of environmental law. During the last two years a nearly permanent dialogue developed between the Commission and the member states to discuss a large variety of issues that resulted from the implementation of the EU ETS directive and the practicalities of emissions trading: i.e., the allocation issues, definitions of installations, the monitoring, reporting and verification of emissions, the register used for the transfer of allowances as well as the financial aspects of emissions trading.

---

[25] Explanatory Memorandum to the Dutch Greenhouse Gas Emission Allowance Trading Act, Parliamentary Documents II, 2003-2004, 29565, No. 3, p. 16: at <http://www.vrom.nl>.

[26] Convention on public access to information, public participation and access to justice in environmental matters (Bulletin of Treaties 2001, No. 73).

The wide range of different consultation processes, sometimes weekly meetings and discussions between the services of the Commission and the member states has thoroughly changed and intensified the relationship and cooperation between the Commission, the member states and other stakeholders. For instance, when the evaluation process for the Commission Decision on the Monitoring and Reporting Guidelines started and the revised guidelines for the second period were being drafted in 2006, member states participated fully and extensively in the drafting process. This was a major improvement compared to the experience with drafting the Monitoring and Reporting Guidelines for the first trading period. This semi-permanent dialogue in Brussels also had a beneficial impact on the relationship and cooperation between member states. Over the last four years they have learned to share experiences and to draw lessons from the best practices in the member states. Furthermore, a range of projects were set up to consult and advise each other through the IMPEL EU-ETS Working Group.[27] This showed that networks such as IMPEL can play an important role in exchanging information and best practices and help to create a level playing field.

6. CONCLUSIONS

Although initially member states did not realize the importance of a harmonized implementation of emission trading, the experiences of the last two years show that they have come a long way in discovering the full consequences of emissions trading and the need for better cooperation and harmonization between the Commission and the member states. A harmonized implementation of emission trading is clearly a prerequisite for the success of the scheme. Uniformity and harmonization are needed in all elements of the compliance chain. The European Commission and member states should therefore agree on common and more precisely formulated definitions as well as on more uniform procedures and requirements. The question then becomes how this commonality could best be achieved. A likely solution will be to empower the Commission in the directive to draw up a regulation for those elements of the directive that need full harmonization and uniform implementation between the member states. The advantage of a regulation is its direct, legally binding effect on the subjects it aims to regulate. In this case operators and other regulated subjects like verifiers and laboratories will be directly affected and regulated. Compared to a Commission Decision or a Directive that needs to be implemented into national law a Commission Regulation would allow a more harmonized ap-

---

[27] The European Union Network for the Implementation and Enforcement of Environmental Law (IMPEL) is an informal Network of the environmental authorities of the member states, acceding and candidate countries of the European Union and Norway. The Network's objective is to create the necessary impetus in the European Community to make progress on ensuring a more effective application of environmental legislation. It provides a framework for policy makers, environmental inspectors and enforcement officers to exchange ideas, and encourages the development of enforcement structures and best practices. At <http://ec.europa.eu/environment/impel/>.

proach, and would bring the clarity and stringency needed for a strict compliance chain.[28]

Furthermore, a harmonized and uniform approach towards all elements of the compliance chain for the post-Kyoto regime should not only be met by legislative means. Other tools like new information technologies are equally important to create a robust and strict compliance mechanism and to ensure a level playing field by structuring the compliance processes and the workflow between the parties involved. In fact, ensuring compliance is basically an issue of exchanging information between the regulator and the regulated subjects. In order to be effective, such information must contain the right compliance content with the right detail provided at the right time to the right addressee. Moreover, information on compliance can only be shared effectively if the three basic principles of an effective emission trading scheme are being met, i.e., transparency, accountability and integrity. To serve these objectives, new information technologies have become indispensable, and IT is certainly to play a much more important role in the third trading period. Automation of the workflow between competent authority, operator and verifier as well as using IT in internal control and verification processes will not only enhance the quality of the data, but will also enable a uniform implementation and enforcement of emissions trading in the member states. This will increase the integrity of the emissions data, the accountability of operators and the transparency with which data are being produced and reported within the EU ETS, thereby strengthening the public's confidence in the market.

It is clear that the revision process of the EU ETS directive must lead to a range of substantial improvements to ensure the more robust emissions trading system needed to be in operation in the third, more ambitious trading period. The main task of the review process is therefore to scrutinize all these elements, e.g., the structures, the procedures and the requirements for the allocation process as well as to examine carefully all the elements of the compliance chain so as to enable the Commission and the member states to realize the very ambitious 2020 greenhouse gas emission reduction targets.

---

[28] See *supra* n. 19.

THE SÃO PAULO PROPOSAL FOR AN AGREEMENT ON FUTURE INTERNATIONAL CLIMATE POLICY

Erik Haites*

ABSTRACT

The São Paulo Proposal aims to create a coherent and balanced package of proposals to catalyze discussions on a post-2012 framework. It outlines a stable, long-term, universal regime based on the principles of equity and common but differentiated responsibilities and respective capabilities. Such a regime is required to encourage the technological change and structural shifts necessary to stabilize greenhouse gas concentrations. Richer countries are expected to adopt binding targets that become more stringent over time. Financial and institutional provisions to enhance developing country implementation of adaptation and to achieve sustainable development are strengthened. Over time they 'graduate' to binding commitments based on their individual circumstances. Adaptation and technology are given more prominent roles. Specified emissions – such as those from international bunkers, a specific industry or a sub-national region of a non-Party – can be addressed through specific agreements. Parties would have the option to impose trade sanctions on non-Parties.

* Margaree Consultants. The provisions proposed in this paper have been put forward by the BASIC Task 4 Team to advance future climate policy discussions and do not express the views or opinions of the funders or the BASIC Project Team as a whole. The Task 4 Team consisted of Gylvan Meira Filho, Institute for Advanced Studies, University of São Paulo, José Goldemberg, Instituto de Eletrotécnica e Energia, University of São Paulo, Jacques Marcovitch, Faculty of Economics, University of São Paulo with support from Erik Haites, Margaree Consultants, Niklas Höhne, Ecofys and Farhana Yamin, Institute of Development Studies, University of Sussex, UK.
The BASIC Project – funded by the European Commission – supported the institutional capacity of Brazil, India, China and South Africa to undertake analytical work to determine what kind of climate change actions best fit within their current and future national circumstances, interests and priorities. Additional funding was kindly provided by the UK Department for Environment, Food and Rural Affairs and Australian the Greenhouse Office. For further information about BASIC, go to <http://www.basic-project.net>.
Earlier versions of this proposal were presented and discussed at the BASIC workshop hosted by the Institute of Advanced Studies, University of São Paulo, in August 2006; a side event in Nairobi, November 2006; the Center for Clean Air Policy (CCAP) Future Action Dialogue in Rio de Janeiro, January 2007; the ECP/BASIC High-level Colloquium, Brussels, March 2007; and the T.M.C. Asser Institute workshop on Tackling Climate Change: An Appraisal of the Kyoto Protocol and Options for the Future, in The Hague, March 2007. The Team acknowledges the constructive comments provided by participants at these events, staff of the Environment Directorate General of the European Commission, James Bruce, Jan Corfee-Morlot, Joanna Depledge, Christiana Figueres, Habiba Gitay, Michael Grubb, Frank Jotzo, Maria Netto, Jim Penman, Tahar Hadj Sadok, Jake Schmidt and Dennis Tirpak. This does not imply support for the São Paulo Proposal by any of these individuals or organizations.

W.Th. Douma et al., eds., The Kyoto Protocol and Beyond
© *2007, T·M·C·ASSER PRESS, The Hague, The Netherlands and the Authors*

1. INTRODUCTION

The United Nations Framework Convention on Climate Change (UNFCCC) has as its ultimate objective the 'stabilization of greenhouse gas concentrations in the atmosphere at a level that would prevent dangerous anthropogenic interference with the climate system.' It also provides that developed country (Annex I) Parties should take the lead in combating climate change and the adverse effects thereof based on the principles of equity and common but differentiated responsibilities and respective capabilities.

To stabilize atmospheric concentrations at any level will require a significant reduction of current global emissions. As a first step, the Kyoto Protocol limits the emissions of developed country (Annex B) Parties during the period 2008-2012. Some Annex I Parties to the UNFCCC, notably Australia and the United States, have not ratified the Kyoto Protocol.

The emission reductions needed to stabilize atmospheric concentrations can only be achieved through sustained efforts over several decades. The technological change and structural shifts needed require a comprehensive, long-term regime. But scientific knowledge, technology and the circumstances of individual countries will change significantly in unforeseen ways over this period.

Discussions on reducing greenhouse gas emissions after 2012 are just beginning. The São Paulo Proposal aims to present a coherent package that is politically acceptable to a wide range of Parties as a substantive input to those discussions. Presenting a coherent package is intended to advance discussion beyond the suite of ideas and options that currently abound in the literature.[1] Each element of the Sao Paulo Proposal is accompanied by an explanation that indicates why it has been included and its salient implications.

The São Paulo Proposal reflects the principles of equity and common but differentiated responsibilities and respective capabilities. All Parties have common responsibilities, but the effective date, and legal nature, of the responsibilities are differentiated based on their capabilities. Richer (Annex I/B) Parties are expected to continue their leadership by using their economic and intellectual resources to reduce their emissions, develop new technologies, and provide a stable stream of funding for adaptation by vulnerable developing country (Non-Annex I) Parties. Non-Annex I Parties are expected to adopt sustainable development policies. Over time, based on their individual circumstances, developing countries are expected to take on emissions limitation commitments.

The São Paulo Proposal is presented here in the form of an amended Kyoto Protocol. The advantage of amending the Kyoto Protocol is that it allows improvements and innovations to be adopted with less risk of reopening what has already been agreed or disturbing established institutional arrangements. The major disadvantages are that non-Parties to the Protocol can participate in the negotiation of

---

[1] See Bodansky, 2004: Gupta, Tirpak, et al., 2007; Kameyama, 2004; and Philibert, 2005 for summaries of proposals.

amendments only as observers and would have to become parties to the Kyoto Protocol to be part of the new regime. Other options for putting the Proposal into legal effect are discussed later.

The rest of the paper presents and explains the elements of the São Paulo Proposal.

1.1 Medium and long-term goals

Parties agree on medium and long-term goals which are used to assess progress toward the ultimate objective of avoiding dangerous anthropogenic interference with the climate system.

Explanation

The adoption of medium and long-term goals enables the Protocol to be more closely tied to climate science and sustainable development. Examples of possible goals include:

(a) a maximum temperature increase of 2°C by 2100;
(b) a maximum atmospheric concentration of CO_2 of 500 ppmv by 2050;
(c) greenhouse gas emissions by Annex I Parties to the Convention at least 50% below their combined 1990 emissions in 2050;
(d) global food supply sufficient to reduce hunger by [X] by [date];
(e) maximum loss of natural ecosystems of [X] by [date].

These goals have no direct consequences for Parties' commitments; rather they provide a basis for the Conference of the Parties serving as the Meeting of the Parties (COP/MOP) to assess progress during the five-year reviews.

1.2 Commitments of Annex I/B Parties

Parties negotiate the annual emission limitation commitments of each Annex I Party for each calendar year from 2013 through to 2018 in tCO_2e/year. The commitments are to include the net emissions due to all land use, land-use change and forestry (LULUCF) activities on all lands within the Party's boundaries.

After the commitments have been politically agreed upon each Party may choose to convert its commitment into a combination of the following legal forms:

– an absolute emissions limit (tCO_2e/year);
– an emissions intensity limit (tCO_2e/unit GDP in constant national currency units); and
– new and additional funding (USD per year) to a maximum of 10% of its commitment.

The rules for conversion and extension of the intensity and financial components are designed to be as stringent as those for the absolute component.

Over time countries that are currently not Annex I Parties would also negotiate such commitments. After a Non-Annex I Party has politically negotiated a commitment that can be approved by the COP/MOP, Annex B is amended to show its commitment for the initial six calendar years.

An Annex I/B Party may request COP/MOP approval for a change to the type and/or level of its commitment.

Explanation

Each Annex I/B Party has a choice of expressing its annual commitments as any combination of absolute and GDP intensity emissions limits[2] and, up to 10%, financial payments.[3] For example, a Party could decide to convert its annual commitments to 50% absolute, 40% intensity and 10% financial.[4] The intensity and financial components are intended to be as stringent as the absolute component over time.[5] For ease of comparison and transparency the annual commitments for 2013-2018 would be *politically negotiated* on the basis of absolute emissions. After agreement is reached, the legal form of the commitments would be chosen by the Party concerned and then set out in the amended Protocol. The initial commitments are for each *calendar year* for the period 2013 through to 2018. With unlimited carryover and assessment of compliance at five-year intervals (as proposed), this provides the same flexibility provided by Kyoto's five-year commitment period.

The net emissions due to all land use, land-use change and forestry (LULUCF) activities on all lands within the Party's boundaries are included in the commitments. This differs from the Kyoto Protocol which allows each Party to choose

---

[2] Ellerman and Wing, 2003, p. 5 notes that linear combinations of an absolute and an intensity target are possible. The intensity coefficient (tCO2e/unit GDP in constant national currency units) would be calculated by dividing the Party's annual commitment for 2013 by its GDP (in national currency units) for 2013 when this figure is reported. Thereafter, the intensity coefficient declines by 3% per year through 2018. When assessing compliance, the intensity coefficient for a given year, say 2017, would be multiplied by the Party's actual GDP for that year (in constant national currency units) to get the allowable emissions for that component of the commitment. Possibilities for 'gaming' – a Party choosing less stringent forms of commitments for itself – are limited.

[3] A decision by the COP/MOP would establish the value per tCO2e to be used to establish the annual emissions equivalent of the financial payment in 2013. The value should be roughly equal to the current international market price for compliance units. The financial payment is increased by the Party's GDP implicit price index each year through 2018. If a Party chooses a financial component, say 5%, for its commitment, compliance with the emissions component would be assessed on the basis of the remaining fraction (95%) of its actual emissions.

[4] Jotzo and Pezzy, 2005, Table V suggest that absolute commitments would be best for Australia, Canada and New Zealand while intensity commitments would be best for Europe, Japan, Russia and the United States.

[5] Intensity targets are sometimes seen as inherently less environmentally stringent than absolute targets. Herzog, et al., 2006 indicates that the form of the target does not determine its environmental effectiveness.

LULUCF activities and/or areas for LULUCF activities that count toward meeting its commitment.[6]

When Annex B is amended to include commitments for another Party or an Annex I/B Party requests a change to the type or level of its commitment, the request would be considered at the subsequent session of the COP/MOP. Adoption would require approval by three fourths of the Parties present and voting. Existing commitments, if any, remain in effect while the request is considered.

Any financial payments would be divided between the Adaptation Fund and the new Technology Funding Mechanism as decided by three fourths of the Parties present and voting at a COP/MOP.

1.3 Automatic extension of Annex I/B commitments

The commitments of all Annex I/B Parties after 2018 would automatically be extended annually rather than be politically negotiated. Commitments for 2019 would be set in 2013, 2020 in 2014, etc., based on the formula agreed during the negotiations and set out in the amended Protocol.

The proposed formula specifies that if conditions during the previous year indicate that compliance for *Annex I/B Parties as a whole* has not been burdensome, the commitments for the next year are made more stringent by the equivalent of 1% for absolute commitments otherwise the stringency of the commitments for the next year remains unchanged. The intensity and financial components of the commitments are adjusted to maintain equivalent stringency.

Specifically, the commitments of all Annex I/B Parties for the next year will be made more stringent if *either* of the following conditions are met:

– the total quantity of compliance units in all registries carried over has increased from the same date during the year prior to the COP/MOP; or
– the international price of AAUs during the year prior to the COP/MOP has not increased by more than the rate of inflation.

Explanation

The automatic extension procedure is a key feature of the São Paulo Proposal. It provides economic and legal certainty far beyond the agreed targets for 2013-2018 whilst still giving Parties a high degree of assurance that the adjustments will not be burdensome. The extension procedure avoids the uncertainty created by commitments that 'expire' every five years. Such uncertainty hampers the long-term investments, structural changes, technological development, and stable carbon markets needed to reduce emissions significantly.

The extension procedure ensures commitments are always known five years in advance and are predictable within a relatively narrow range (equivalent to 0 to -

---

[6] Arts. 3.3 and 3.4 would be deleted and Annex A of the Kyoto Protocol (coverage) would be amended to include LULUCF activities.

10% for absolute commitments) for the following ten years.[7] Adjustments to commitments are small (equivalent to 1% for absolute commitments) and occur only if compliance is not burdensome. This approach is more attuned to the 10-40 year investments typical of major carbon emitting sectors such as power generation, transport and industry.

Commitments are made more stringent only if the trigger conditions demonstrate that compliance for Annex I/B Parties *as a group* has become easier or less costly during the previous year. The commitments of all Annex I/B Parties for the next year will be made more stringent if *either* of the following conditions are met:[8]

- the total quantity of compliance units in all registries carried over has increased from the same date during the year prior to the COP/MOP; or
- the international price of AAUs during the year prior to the COP/MOP has not increased by more than inflation.

If compliance does *not* become easier or less costly (neither condition has been met):[9]

- the absolute component of the commitment remains unchanged for the next year;
- the intensity coefficient is reduced by 3%; and
- the financial component is increased by the Party's GDP implicit price index.

If compliance *is* becoming easier or less costly (one of the conditions has been met):

- the absolute component of the commitment is reduced by 1% for the next year;
- the intensity coefficient is reduced by 5%; and
- the financial component is increased by the increase in the Party's nominal GDP.[10]

The extension procedure applies to the commitments of all Annex I/B Parties, i.e., if either of the trigger conditions are met, all Annex I/B Parties have their commitments made more stringent automatically by the same amount at the same time.

---

[7] The five year reviews could lead to changes that affect the latter part of this period.

[8] These conditions would involve agreeing international procedures to determine the total quantity of compliance units in all registries carried over and the change in the international price of AAUs. Based on these procedures, assessment of whether the trigger conditions have been met would be undertaken by a technical, independent body such as the Enforcement Branch of the Compliance Committee.

[9] Analyses reported in papers on the São Paulo Proposal available on the BASIC website compare absolute and intensity targets for some Annex I/B Parties over different time periods. These analyses suggest that the proposed adjustments to intensity coefficients yield commitments of comparable stringency to the corresponding absolute commitments.

[10] In other words, the financial component is adjusted for both inflation and real growth.

This provides certainty that all are pulling together in the same direction under the same constraints even if their actual targets have a different legal form.

It is unlikely that a single Annex I/B Party could influence the trigger conditions. Even if this did happen, it would affect all Annex I/B Parties equally. In other words, a Party that tried to manipulate the trigger conditions would not gain a competitive advantage over other Annex I/B Parties.

Notwithstanding the extension procedure, a Party would be able to request an amendment to Annex B to change the type or level of its commitment at any time.

1.4   Economic hardship

An Annex I/B Party whose real GDP has declined by more than 1% during a calendar year may request a compliance exemption for that year. Then its commitment for the year would be deemed to be equal to its actual emissions during the year.

Explanation

The economic hardship provision addresses concerns associated with the economic risks of long-term commitments. An *individual* Annex I/B Party can request a compliance exemption for a calendar year if its real GDP has declined by more than 1% during that year. Then its commitment for the year will be deemed to be equal to its actual emissions during the year.[11] This means that it will not bear an additional economic burden due to meeting its emissions limitation commitment for that year. The number of consecutive exemptions is not limited.[12]

1.5   Non-Annex I Parties' quantified sustainable development actions and 'no lose' commitments

The stronger commitments by Annex I/B Parties are accompanied by a wider range of actions by developing countries to reduce their emissions in ways that enhance sustainable development. A Non-Annex I Party may:

– host Clean Development Mechanism (CDM), including programmatic CDM, projects;
– quantify the emission reductions achieved by its sustainable development actions, including policies to reduce deforestation; and

---

[11] A Party would be required to notify the Enforcement Branch of the Compliance Committee of its wish to use this provision for a given year before compliance for that year is assessed. The provision is invoked *ex post*; if, for example, publication in 2019 of a Party's economic data for the year 2018 indicate a decline in its real GDP of more than 1% during 2018, the Party may notify the Enforcement Branch that it wishes to have its actual commitment for 2018 deemed to be equal to its actual emissions during 2018.

[12] This provision also ensures that a Party suffering economic hardship stays within the regime rather than withdrawing and threatening the long-term stability of the regime.

– adopt a sectoral, excluding LULUCF, or national 'no lose' commitment.[13]

A Non-Annex I Party may adopt any combination of these options as long as it does not lead to double counting of emission reductions.

Non-Annex I Parties that volunteer to quantify the emission reductions achieved by their sustainable development action must calculate the emission reductions achieved using methodologies agreed by the COP/MOP and report them regularly through their national communications. Quantified sustainable development actions under this option can *not* generate tradable credits. But Parties that quantify their emission reductions would be entitled to use simplified procedures to access funding from the Adaptation Fund and proposed Technology Funding Mechanism (see Elements 12 and 13 below)

Non-Annex I Parties may also adopt *a sectoral or national 'no lose' commitment* which can earn voluntary emission reduction units (VERs) for the net emission reductions achieved. VERs (equivalent to CERs for Annex I/B compliance purposes) can be banked or traded. A proposed commitment would be subject to independent review by the CDM Executive Board to ensure that it is more stringent than the emissions that would otherwise occur and assess the calculation of net emission reductions achieved. A proposed commitment must be approved by three fourths of the Parties present and voting at a session of the COP/MOP and be maintained until the Party becomes an Annex I/B Party and adopts a binding national commitment.

Explanation

The São Paulo Proposal gives effect to the principle of common but differentiated responsibilities by allowing developing countries to accelerate actions to reduce emissions consistent with their sustainable development priorities without taking on legally binding commitments.

Reductions achieved by sustainable development actions
A Non-Annex I Party may elect to quantify and report the emission reductions achieved by its sustainable development actions. At present, although all Parties have committed themselves to implementing and publishing measures to mitigate and adapt to climate change, there is no clear institutional mechanism to measure and report the resulting emission reductions.[14] As a result the very significant contribution of developing countries to global climate protection is insufficiently acknowledged. Being able to report the reductions achieved by sustainable development

---

[13] A 'no lose' commitment (also called 'non-binding' or 'one way' commitments) entails no compliance obligation if actual emissions exceed the commitment, but allows the surplus credits to be sold if actual emissions are lower than the commitment.

[14] Formulation, implementation, publication and regular updating of national (and where appropriate regional) programmes containing measures that mitigate and adapt to climate change is mandatory for all Parties under Art. 4.1(b) of the UNFCCC and Art. 10(b) of the Kyoto Protocol.

actions addresses this gap and allows the contributions of developing countries to be counted and politically recognized. Quantified sustainable development actions do not generate tradable credits. A Non-Annex I Party that prefers to generate tradable credits can implement its sustainable development actions as programmatic CDM projects or adopt a 'no lose' commitment that generates VERs.

The benefits of electing to quantify and report the emission reductions achieved by sustainable development actions are:

- international recognition for the emission reductions achieved;
- use of simplified procedures to access funding from the Adaptation Fund and Technology Funding Mechanism; and[15]
- by reducing per capita emissions, deferral of the date when the Party reaches its cap on transfers of CERs and VERs (see Element 6 below).

Quantified sustainable development actions by Non-Annex I Parties would not affect the existing UNFCCC commitment for Annex II Parties to provide new and additional funding for matters covered by Convention Articles 4.1, 4.4 and 4.5.

Methodologies for calculating emission reductions due to the implementation of sustainable development actions could be developed by an existing body such as the Consultative Group of Experts on National Communications from Non-Annex I Parties (CGE) for approval by the COP/MOP.[16] Results and emission reductions would be reported in national communications.[17] The UNFCCC secretariat could synthesize the information from national communications and report periodically to the COP/MOP.

'No lose' commitments
Provided there is no double counting with other options, a Non-Annex I Party may also adopt a sectoral (such as electricity generation) or national 'no lose' commitment. A commitment may take the form of an absolute limit or a limit on the emissions per unit of output for the sources covered. A proposed commitment must be reviewed by the CDM Executive Board to ensure that the:[18]

- commitment is more stringent than the emissions that would otherwise occur;
- methodology for monitoring the actual emissions of sources covered by the commitment is appropriate;

---

[15] The entitlement is to (i) fast tracking of funding requests and (ii) for these to be on the basis of simplified procedures.

[16] The CGE's mandate would need adjustment to allow this.

[17] Reporting the achievements of quantified sustainable development actions would create more legal certainty about the regularity of national communications without altering the legal provisions in the Convention that these must be financed by Annex II Parties.

[18] The CDM Executive Board would adopt a procedure for reviewing proposed commitments. The procedure could include review by the Meth Panel and an accredited Designated Operational Entity as in the case of CDM projects, or a new procedure involving different experts.

- possible increases in emissions by other sources and double-counting of reductions claimed by CDM projects are identified and incorporated into the calculation procedure; and
- procedure for calculating the net emission reduction achieved by the commitment is reasonable.

Given a recommendation for approval by the CDM Executive Board, a proposed 'no lose' commitment would be approved by three fourths of the Parties present and voting at a session of the COP/MOP.[19] Once the COP/MOP had approved a 'no lose' commitment, it would be treated as registered by the CDM Executive Board.

The Non-Annex I Party government with a 'no lose' commitment periodically would retain an accredited Designated Operational Entity to verify and certify the net emission reductions achieved since the previous verification. The CDM Executive Board would issue voluntary emission reduction units (VERs) for the certified emission reductions into the Party's account in the CDM registry. VERs could be traded or banked and would be equivalent to CERs for Annex I/B compliance purposes. Like CERs, VERs would be subject to the 2% levy for the Adaptation Fund. To avoid instability, a 'no lose' commitment would need to be maintained until the Party becomes an Annex I/B Party with a national emissions limitation commitment.

1.6   Graduation by Non-Annex I Parties

A Non-Annex I Party is expected to become an Annex I/B Party and adopt a national emissions limitation commitment when its cumulative transfers of CERs and VERs since 1 January 2005 reach its share of the global cap on such transfers.[20] The Proposal is that each Non-Annex I Party's share of the global cap is based on its population and an index that reflects its responsibility, capability, and potential to mitigate. The shares, and hence the transfer limit for each Party, would be recalculated at five-year intervals to reflect changing developing country circumstances. The global cap and formula for calculating the share for each Non-Annex I Party would be agreed as part of the political post-2012 negotiations.

Once the cumulative transfers of CERs and VERs since 1 January 2005 by a Non-Annex I Party reach its cap, the Party is expected to become an Annex I/B Party and accept a national emissions limitation commitment, bearing in mind that this commitment is negotiated and so can reflect national circumstances and that the Party may choose the form of its target: absolute, intensity or financial.[21] A

---

[19] The role of the CDM Executive Board is limited to recommending approval (or rejection) of proposed 'no lose' commitments to the COP/MOP.

[20] If additional compliance units that can be generated by Non-Annex I Parties are created they would be included in the calculation as well. For example, if a new mechanism is agreed for generating credits for reduced deforestation cumulative transfers of those units would also be counted against the limit.

[21] Annex B would be amended to list the first six years of the commitment proposed by such a Party and approved by three-fourths of the Parties present and voting at a session of the COP/MOP.

Non-Annex I Party that does not adopt a national emissions limitation commitment when it has reached its cap is deemed to have withdrawn from the Protocol.

Explanation

A stable, universal regime to address climate change must include a provision for determining when a Non-Annex I Party is expected to adopt an emissions limitation commitment.[22] In the São Paulo Proposal this mechanism is a cap on the cumulative transfers of CERs and VERs by each Non-Annex I Party.[23] This means that a Non-Annex I Party 'graduates' when it meets agreed criteria, rather than on an arbitrary date.

A global cap on transfers of CERs and VERs would be shared among Non-Annex I Parties based on each Party's population multiplied by an index that reflects three factors: responsibility, capability, and the potential to mitigate.[24] Responsibility is quantified as cumulative CO_2 emissions per capita since 1990, capability as GDP per capita for the most recent year available, and potential to mitigate as total greenhouse gas emissions, excluding emissions from land-use change and forestry, per capita for the most recent year available. Higher cumulative emissions per capita, GDP per capita and GHG emissions per capita yield a lower value for the index and hence a smaller share of the global cap. A larger population produces a higher share of the cap.

The limits on transfers of CERs and VERs determine when a Non-Annex I Party graduates based on its individual circumstances in an equitable and balanced manner that:

- ensures that Annex I/B Parties undertake significant emission limitation commitments (including purchases of CERs and VERs) *before* Non-Annex I Parties are expected to adopt commitments. Weak commitments by Annex I/B Parties reduce the quantity of CERs and VERs they purchase, which delays the date when each Non-Annex I Party reaches its cap. Conversely, stronger commitments by Annex I/B Parties advance the dates when Non-Annex I Parties accept binding obligations;
- allows Non-Annex I Parties to benefit from participation in the carbon market and receive increased funding for adaptation and technology development *before* being expected to adopt binding emission limitation commitments;

---

[22] A system of short, 5 year, commitments that are periodically renegotiated, implicitly assumes that groups of Non-Annex I Parties adopt commitments as part of each renegotiation.

[23] CERs would include lCERs and the maximum number of tCERs transferred during any five year period.

[24] By way of example, a global limit of 20 billion tCO2e is likely to mean a lag between adoption of commitments by Annex I/B and Non-Annex I Parties of 16 to 40 years, much longer than the 10 year grace period for developing countries provided in the amended 1987 Montreal Protocol on Substance the Deplete the Ozone Layer for developing countries. Of course, the lag would be shorter for some Non-Annex I Parties and much longer for others.

- creates an incentive for every Non-Annex I Party to pursue a less emissions-intensive development path, even if it does not earn CERs or VERs, since that increases its cap and its benefits from participation in carbon markets. A Non-Annex I Party also could earn CERs or VERs for reducing its emissions, but retains some of those units to delay the date at which it reaches its limit and to help it comply with its subsequent commitment;
- provides an equitable geographic distribution of the benefits of participating in the Protocol in the long run. As individual Non-Annex I Parties graduate, the share of CERs and VERs which the remaining Non-Annex I Parties can supply to carbon markets is increased;
- recognizes the changing circumstances – population, per capita GDP, per capita GHG emissions – of individual Non-Annex I Parties over time.

It is important to remember that CERs and VERs do not reduce net emissions to the atmosphere. The emissions reduction in the Non-Annex I Party is offset by the emissions of an Annex I/B Party. Both Parties benefit economically, but there is no climate change benefit. Ultimately, therefore, transfers of CERs and VERs must be limited and Non-Annex I Parties must adopt emission reduction commitments.

Procedurally, the size of the global cap and an agreed process for how this cap would be shared among Non-Annex I Parties and be periodically recalculated would be negotiated. Responsibility for implementing the process would be given to the Facilitative Branch of the Compliance Committee. The data would be obtained from Non-Annex I national communications.

A Non-Annex I Party that did not adopt a quantified commitment when it reached its transfer limit would be deemed to have withdrawn from the Protocol (although it would remain a Party to UNFCCC).[25] Withdrawal from the Protocol would mean loss of access to benefits such as transfers of CERs or VERs, access to the Adaptation Fund and the Technology Funding Mechanism, and the possible imposition of trade restrictions (see element 15 below).

1.7   Clean Development Mechanism

The Clean Development Mechanism continues to function with the following refinements:

- afforestation and reforestation project activities will continue to be eligible; and
- project activities that reduce emissions of gases other than CO_2 will be limited to a single crediting period and be eligible only if the host government requires the measures implemented to remain in operation after the end of the crediting period.

---

[25] The COP/MOP would need to adopt a procedure for such a determination and implementation of the consequences.

If a Non-Annex I Party does not ratify the amended Protocol by 30 September 2012, CERs, lERs and tCERs held in accounts of the Party and legal entities approved by the Party would be ineligible for transfer to other accounts after 31 December 2012.

Explanation

The CDM is of fundamental importance to developing countries and an important element of Annex I/B Parties' compliance efforts. The São Paulo Proposal is that the provisions of Article 12 of the Kyoto Protocol, all decisions of the COP and COP/MOP relating to the CDM, and all decisions of the Executive Board remain in effect.

Currently, a small number of large projects involving HFC-23 and N_2O destruction with few sustainable development benefits are crowding out the CDM market. The Proposal to limit CDM projects that reduce gases other than CO_2 to one crediting period would provide more support for renewable energy and energy efficiency projects and implicitly widen the geographic benefits of the CDM – to the likely benefit of LDCs and smaller countries. Afforestation and reforestation projects, currently limited to the Kyoto Protocol's first commitment period, remain eligible indefinitely, again benefiting a wider range of Non-Annex I Parties.

To provide stability and security for the carbon markets it is important that there is no gap between the end of the Kyoto Protocol and the start of the new regime. To create an incentive to ratify the amended Protocol in time for it to enter into force on 1 January 2013 and so provide economic security for existing CDM investments, CERs, lERs and tCERs in the accounts of a Non-Annex I Party and those of its approved legal entities could not be transferred after 31 December 2012 if it did not ratify the revised Protocol by 30 September 2012. Similar provisions would apply to ERUs, RMUs and AAUs to encourage Annex I/B Parties to ratify the revised Protocol by 30 September 2012 as well.

A transitional issue that arises when a Non-Annex I Party adopts quantified emissions limitation commitments is the fate of the CERs, lCERs and tCERs that could be issued for reductions achieved prior to the end of the current crediting period for each project registered prior to the effective date of the commitment.[26] The São Paulo Proposal is that each existing project continues to earn CERs, lCERs and tCERs for verified emission reductions for the balance of its current crediting period and that those units should *not* be deducted from the host country's national commitment.[27] Since the emission reductions achieved by those projects help the Party to achieve its newly adopted commitment this would create some double counting, but it has the benefit of providing security for existing CDM investments.

---

[26] A 'no lose' commitment adopted by a Non-Annex I Party is replaced by its national emissions limitation commitment when it becomes an Annex I/B Party, so VERs can no longer be issued.

[27] The crediting period could not be renewed. At the end of the crediting period the host country could decide to allow the project to register as a Joint Implementation project.

1.8   Joint Implementation

Joint Implementation would continue to function unchanged except that ERUs transferred from the national registry of the host Party would be subject to a share of proceeds equal to 2% to assist technology development.

Explanation

The São Paulo Proposal carries forward the provisions of Article 6 of the Kyoto Protocol, all decisions of the COP and COP/MOP relating to Joint Implementation, and all decisions of the Joint Implementation Supervisory Committee into the amended Protocol.

The Proposal provides that a 2% share of proceeds be applied to any transfer of ERUs out of the national registry where they are issued. This is a one time tax unless the unit is sent back to its initial registry and then exported a second time. This share of proceeds would go to the Technology Funding Mechanism.

1.9   Emissions trading

Emissions trading would continue to function unchanged under the São Paulo Proposal except that:

– AAUs and RMUs transferred from the national registry into which they were issued will be subject to a share of proceeds equal to 2% to assist technology development; and
– carryovers of AAUs, CERs, tCERs, lCERs, ERUs, RMUs or VERs will not be restricted.

Explanation

Emissions trading is a crucial part of the system of global cooperation to lower the costs of compliance. The provisions of Article 17 of the Kyoto Protocol and all decisions of the COP and COP/MOP relating to Emissions Trading are reaffirmed and given legal effect under the São Paulo Proposal.

The 2% share of proceeds is applied to any transfer out of the national registry where an AAU or RMU was issued. This would be a one time tax on units traded internationally as traders would avoid transferring units back into the national registry where the units were issued and out a second time.

A further improvement is the removal of the current restrictions on banking or carryovers of units. These restrictions are mainly of an historic interest as in practice they are readily circumvented.

1.10   Compliance

Although the quantified emission limitation commitments of Annex I/B Parties apply to specified calendar years, compliance is determined at five year intervals as in the case of the Kyoto Protocol. In cases of non-compliance, any AAUs to be deducted from a Party's future allocation would be deducted from the allocation for the year after the decision is made by the Compliance Committee.

Explanation

Current compliance procedures and mechanisms are fundamental to the integrity of the Protocol and the functioning of carbon markets. Accordingly, the provisions of Article 18 of the Kyoto Protocol, all decisions of the COP and COP/MOP relating to Article 18, and all decisions of the Compliance Committee remain in effect until the entry into force of the amended Protocol. The compliance mechanisms and procedures would be included in the amendments to incorporate them into the new regime in a legally binding form.

Additional modalities for the Enforcement and Facilitative Branches to cover the functions set out in the Proposal would also need to be agreed. The COP/MOP would, for example, need to decide whether the date for determining compliance is the same for all Annex B Parties (for example, 2013-2018 for all) or differs by Party (2013-2016 followed by 2017-2021 for some; 2013-2017, followed by 2018-2022 for others; etc.). The COP/MOP would also need to decide how to deal with the non-payment of the financial component of a Party's commitment.[28]

1.11   Enhanced implementation of adaptation

A permanent Adaptation Committee of Experts (ACE) is established to provide advice on adaptation activities and funding. ACE would also act as a focal point for institutional and policy linkages with international and national bodies charged with the achievement of development goals and with disaster risk reduction and relief.

A five-year pilot phase of 'Adaptation Activities Implemented Cooperatively' is launched by a COP/MOP decision at the time of the adoption of this Protocol through 'prompt start' provisions. The objective of the pilot phase is to catalyze rapid learning about adaptation 'good practices' by supporting enhanced implementation of demonstration projects, programmes and policies in vulnerable countries and communities.

Beginning in the pilot phase, all Parties agree to review, and revise as necessary, design parameters and standards for infrastructure and equipment to incorporate the projected impacts of climate change.

---

[28] The Proposal is that if this is not received in full by 31 March of the following year, the case be referred to the Facilitative Branch of the Compliance Committee. If full payment for all five years has not been received when its compliance is assessed, the Party would be deemed out of compliance.

Parties commit to adopt an appropriate legal instrument to give effect to a risk management or insurance mechanism to address the impacts of extreme weather events by 2010.

Funding for adaptation is enhanced by extending the 2% share of the proceeds on CERs to VERs and contributing proceeds from auctioned allowances for international bunkers and funds from Annex I/B Parties financial commitments to the Adaptation Fund as decided by the COP/MOP (see Element 2).

To ensure the effective targeting of Adaptation Fund resources, based on recommendations by ACE, the COP/MOP would give additional guidance on the use of relevant tools and techniques to assess vulnerability and adaptation options for human populations and natural ecosystems.

Explanation

Adaptation is an immediate, as well as an ongoing, long-term challenge that merits a permanent, institutionalized form of oversight and encouragement and a higher, better-funded profile in the post-2012 regime. Currently, adaptation issues are dealt with in a piecemeal, *ad hoc* manner without proper institutional support. To redress this, a new, permanent Adaptation Committee of Experts (ACE) is established by a COP/MOP decision adopted at the same time as the Protocol without the need for the amended Protocol to enter into force. This is similar to the prompt start of the CDM which was set up in 2001 under the auspices of the COP before the Kyoto Protocol entered into force.

The composition of the ACE could be similar to that of the Consultative Group of Experts on National Communications from Non-Annex I Parties (CGE).[29] ACE would act as a focal point within the Protocol for institutional and policy linkages with international and national bodies charged with the achievement of Millennium Development Goals and with disaster risk reduction and relief, such as International Strategy for Disaster Reduction, and national platforms for disaster risk reductions established under the Hyogo Framework for Disaster Risk Reduction. The LDC Expert Group (LEG), which currently operates under the Convention and whose mandate is limited to NAPA preparation and expires in 2007, would be replaced by a specialist panel of the ACE.

A five-year pilot phase of 'Adaptation Activities Implemented Cooperatively' is launched by a COP/MOP decision at the time the amended Protocol is adopted, thus allowing the pilot phase to start prior to the entry into force of the amended Protocol. ACE's mandate in relation to the pilot phase would be to:

– support the registration of implemented pilot phase projects, programmes and policies, particularly from vulnerable developing countries and communities;
– facilitate funding for adaptation activities from larger sources of development funding such as multilateral, bilateral and private channels;

---

[29] 24 experts drawn from a government-nominated roster of experts with 5 each from Asia, Africa, GRULAC, six from Annex I including one EIT plus 3 experts from international organizations.

- synthesize policy-relevant information from the pilot phase on, inter alia, good practices, relevant information and tools;
- support the development and dissemination of risk management approaches that include responses to climatic extremes, extreme weather events and climate change; and
- support institutional and policy linkages between adaptation to climate change and bodies charged with the achievement of Millennium Development Goals (MDGs), and disaster risk reduction and relief.

The review of design parameters is particularly important given the scale of infrastructure about to be put in place in developing and developed countries over the next 10-20 years. The review would cover design parameters for, *inter alia*, port facilities, sea walls, canals, dams, water systems, irrigation systems, storm and sanitary sewers, residential buildings, commercial and industrial buildings, roads, railways, bridges, communication systems, and electricity grids. For example, the review and revision could specify the sea level, wave height, and other parameters to be used for the design of sea walls in the country taking into account the projected effects of climate change. The revised parameters would need to be incorporated into the relevant legislation, regulations, professional standards, etc., to ensure that they are used in the design of new facilities or changes to existing facilities.

The review would also cover standards for, *inter alia*, residential appliances, commercial and industrial equipment, and vehicles. It would revise the standards to improve energy and water efficiency and account for the projected effects of climate change. The revised standards would need to be incorporated into relevant legislation and regulations by all Parties. Non-Annex I Parties may request funding for the review and revision of design parameters and standards from the Adaptation Fund.

ACE is specifically mandated to develop tools and techniques to assess vulnerability and adaptation options for human populations and natural ecosystems and recommend them to the COP/MOP for adoption with the approval of three-fourths of the Parties present and voting. Once adopted the Adaptation Fund would use these tools and techniques to focus its financial resources primarily on programmatic approaches and projects in developing countries that help ecosystems and people particularly vulnerable to the adverse effects of climate change to adapt to climate change.

Despite the implementation of measures to adapt to climate change, extreme weather events will continue to inflict serious damage, especially to the most vulnerable ecosystems and human populations. The São Paulo Proposal addresses this reality by requiring that Parties agree to assess mechanisms to manage the risk of damage and to provide compensation for damages incurred as a result of extreme weather events and to adopt an appropriate legal instrument to give effect to such mechanism(s) no later than the end of 2010.[30]

---

[30] Emergency response to extreme weather events – medical assistance, food, water, shelter, etc. – is better addressed by institutions such as the International Red Cross/Red Crescent and hence is excluded from this revised Protocol.

1.12 Technology transfer

Technology transfer is a commercial transaction. The São Paulo Proposal believes the amended Protocol can best assist such transactions by enhancing systems to provide information on available technologies and by resolving disputes over technology transfer.

Explanation

The Proposal is to enhance the information available in databases such as TTClear clearinghouse operated by the UNFCCC secretariat and the Climate Technology Initiative of the International Energy Agency, with expert advice available to entities in Non-Annex I Parties.

Some developing countries believe that developed countries restrict the transfer of needed technologies. And some developed countries believe that developing countries impose barriers that restrict the transfer of appropriate technologies. The São Paulo Proposal encourages a Party that believes another Party is restricting the transfer of a technology to present its case to the Facilitative Branch of the Compliance Committee. This is already possible (see Decision 27/CMP.1, Annex, Section XIV Consequences applied by the Facilitative Branch).

1.13 Technology research and development

A Technology Funding Mechanism is established with funds derived from a 2% levy on international transfers of AAUs, ERUs and RMUs and a share of the funding from Annex I/B Parties' financial commitments as decided by the COP/MOP (see Elements 2, 8 and 9). The Technology Funding Mechanism would be managed by an Executive Board operating under the guidance of the COP/MOP.

The main function of the Technology Funding Mechanism would be to consider requests from Non-Annex I Parties for funds to participate in international efforts to develop mitigation and adaptation technologies and to enhance diffusion of relevant technologies by buying down their cost. The Technology Funding Mechanism could also decide to participate in international technology research and development efforts directly.

The Executive Board of the Mechanism would recommend to the COP/MOP how best to use the intellectual property rights acquired by itself or by Non-Annex I Parties as a result of research it has helped to fund.

Explanation

A number of channels to support collaborative research and development already exist under the Convention (the GEF and the Special Climate Change Fund) and through a variety of bilateral and multilateral channels. The São Paulo Proposal aims to enhance the ability of Non-Annex I Parties to participate in such collaborative research and development efforts.

The Executive Board of the Technology Funding Mechanism would replace the Expert Group on Technology Transfer (EGTT) so that one group serves both the Convention and the Protocol. The Executive Board could establish permanent and 'ad hoc' panels to assist with independent screening of funding proposals or to provide advice on specific issues.

The Technology Funding Mechanism would operate mainly in a responsive mode, inviting requests for research funding from Non-Annex I Parties and consortia they wanted to be involved in. Requests would have to indicate how the share of the Party's total cost of participation is to be funded and how ownership of any intellectual property developed by the research programme would be handled. Requests for funding to enhance technology diffusion would have to indicate the anticipated reduction in the cost of the technology and the emissions reduced.

Modalities for the Board and Fund's operation could be modeled on a number of public and private research funds that disburse significant monies for research, development and diffusion at the national and international level. Requests for funding would have to undergo robust screening by independent experts before being considered by the Executive Board. Whilst no procedure can guarantee that the Board will only pick 'winning' technologies, its disbursement modalities would aim to ensure that requests are based on merits and cost effectiveness, rather than political criteria.

1.14   Memoranda of Understanding extending the scope of the agreement with non-Parties

The COP/MOP, with the support of three fourths of the Parties present and voting, may approve a memorandum of understanding (MOU) with national or sub-national government(s) of a country that is not a Party to the Protocol.

Explanation

The São Paulo Proposal is designed to appeal to current non-Parties by, for example, encouraging universal participation, providing more flexible forms of targets and recognizing economic hardship implications. But this may not be sufficient to encourage all UNFCCC Parties to join the revised Protocol in a timely fashion. The provision on MOUs enables, in special cases, the COP/MOP to extend the geographic scope of the revised Protocol. Special situations could include a country that has a difficult ratification process but which is eager to cooperate with the international effort[31] or one or more sub-national governments of a non-Party.[32]

---

[31] In the US ratification of a treaty requires a two-thirds majority in the Senate. However, most international agreements are approved by 'congressional executive agreements' which only require a simple majority in both houses of Congress.

[32] For example, with states that have/will have emissions trading schemes such as California and the states participating in the Regional Greenhouse Gas Initiative (RGGI) or New South Wales, Australia.

1.15 Memoranda of Understanding extending the scope of the agreement to special sectors and sources

The COP/MOP, with the support of three fourths of the Parties present and voting, may approve a memorandum of understanding (MOU) with:

- an entity with legal authority to limit emissions outside the boundaries of Parties (such as international aviation and marine emissions); or
- an entity with legal authority to limit the emissions of specified sources located in more than one Party (such as global emissions of a specific industry).

Explanation

Some sectors may justify special treatment due to legal issues about jurisdiction or concerns about global competitive effects. A MOU offers a mechanism for bringing such sectors into the international framework.

Emissions due to international aviation and shipping could be addressed by negotiating a MOU with a suitable entity, such as ICAO or IATA in the case of international aviation and IMO for international shipping. Each MOU would establish an emissions cap for the sources covered separate from the national commitments of Annex I/B Parties. The emissions cap would be subject to the same automatic extension and adjustment as the national commitments of Annex I/B Parties. Each MOU would specify how allowances equal to the emissions cap would be distributed. The São Paulo Proposal is that an increasing share of the allowances to be auctioned will see the proceeds going to the Adaptation Fund.

Sectoral or industry agreements to limit emissions, such as a global agreement with the aluminum industry, are also possible although they would be more complex due to the need to ensure compatibility with national emissions inventories.

1.16 Trade restrictions

Three-fourths of the Parties present and voting at a session of the COP/MOP may approve trade restrictions to be applied by Parties and countries with approved MOUs against countries that are not a Party to the revised Protocol and that do not have a MOU approved by the COP/MOP.

Explanation

To encourage universality, and as successfully included in several environmental agreements such as the 1987 Montreal Protocol, this provision of the amended Protocol would allow:

- negotiation of a memorandum of understanding (MOU) with a non-Party; and
- regulation of trade with non-Parties in goods and services which if left unregulated could undermine the achievement of the objectives of the Protocol

Countries that refuse to become Parties and do not have an agreed MOU with the COP/MOP could free ride on the efforts of others gaining competitive advantages at the expense of the climate and other Parties. This could be a justifiable reason for the imposition of trade restrictions against that non-Party. Trade restrictions are expected to be applied infrequently, if ever, and to target goods and services that directly undermine the achievement of the objectives of the Protocol, such as fossil fuels. Under a few environmental agreements with such provisions, trade restrictions have been applied against some countries.[33] No instance of a trade restriction under an environmental agreement has yet been appealed to the World Trade Organization.[34]

1.17 Review

To ensure the regime remains responsive to climate science and advances in technology as well as Parties' changing circumstances, the Protocol would be reviewed every five years starting in 2017.

Explanation

The review of the Protocol could lead to the adoption, with the approval of the Parties, of more stringent targets than agreed in the initial negotiations. This happened in the 1987 Montreal Protocol on several occasions when the costs of reductions proved lower than expected and Parties were happy to accelerate their phase-out schedules.

1.18 Legal form of the post-2012 agreement

The São Paulo Proposal is currently presented in the form of an amended Kyoto Protocol. The new elements could be incorporated in accordance with Articles 20 and 21 of the Protocol taking into account the review provisions in Articles 3.9, 9 and 13.

Explanation

A post-2012 agreement could be legally implemented in a number of ways. The advantage of amending the Kyoto Protocol to include the São Paulo elements is that it allows improvements and innovations to be adopted with less risk of reopening what has already been agreed or disturbing the established institutional arrangements (CDM, JI, International Transaction Log, Compliance Committee, Expert

---

[33] Some experts have concluded that the threat or application of trade restrictions has increased participation in some instances.

[34] A non-Party that appeals to the WTO faces the challenge of arguing that countries trying to protect the environment should be punished for trying to induce non-Parties to do their share. Of course the non-Party could still win on legal grounds.

Review Teams, inventories, and reporting). The major disadvantages are that non-Parties to the Protocol can participate in the negotiation of amendments only as observers and would have to become parties to the Kyoto Protocol to be part of the new regime.[35]

The São Paulo Proposal could also be incorporated into a new protocol adopted pursuant to Article 17 of the UNFCCC. A new protocol allows all UNFCCC Parties to participate in the negotiations and it can differ significantly from the Kyoto Protocol. However, a new protocol risks undermining what has been agreed under the Kyoto Protocol and may involve the creation of new institutions that duplicate those established by the Kyoto Protocol until the latter are no longer needed.

A third option is to divide the São Paulo Proposal elements into those that need to be incorporated into a revised Kyoto Protocol and those that can be advanced through actions under the current provisions of the UNFCCC. This would be more complex than an amendment of the Kyoto Protocol or a new protocol.

References

Bodansky, Daniel, 2004. 'International climate efforts beyond 2012: A survey of approaches'. Pew Climate Center, Washington DC.

Ellerman, Denny and Ian Sue Wing, 2003. 'Absolute vs. Intensity-Based Emission Caps,' MIT Joint Program on the Science and Policy of Global Change, Report 100, Massachusetts Institute of Technology, Cambridge, MA., July.

Gupta, Sujata, Dennis A. Tirpak, et al., 2007. 'Policies, Instruments and Co-operative Arrangements', Chapter 13, Working Group III contribution to the Fourth Assessment Report of the Intergovernmental Panel on Climate Change, Cambridge University Press, Cambridge.

Herzog, Timothy, Kevin A. Baumert and Jonathan Pershing, 2006. 'Target: Intensity: An Analysis Of Greenhouse Gas Intensity Targets', World Resources Institute, Washington, D.C., November.

Jotzo, Frank and John C.V. Pezzey, 2005. 'Optimal intensity targets for emissions trading under uncertainty', Economics and Environment Network Working Paper EEN0504, Australian National University, 21 June.

Kameyama, Y., 2004. 'The future climate regime: a regional comparison of proposals', *International Environmental Agreements: Politics, Law and Economics*, 4: 307-326.

Philibert, Cedric, 2005. 'Climate Mitigation: Integrating Approaches for Future International Cooperation', OECD/IEA Annex I Expert Group, Paris.

---

[35] A number of mechanical options are available to enable the USA and Australia to rejoin Kyoto. For example, the proposal to amend Kyoto could include revisions to Annex B to facilitate ratification of the Kyoto Protocol by non-parties at the time of adoption of these amendments.

STAKEHOLDER VIEWS ON APPROACHES AND INSTRUMENTS FOR CONTINUED AND FUTURE CLIMATE CHANGE MITIGATION EFFORTS POST-2012

Robert Tippmann*

1. Introduction

This article addresses some of the main questions related to the continued and future international climate change mitigation efforts in the context of a stakeholder survey conducted in 2007 by Ecosecurities. Views on how climate change should be continued are analyzed by trying to identify common positions, major frontlines with diverging views of stakeholder groups and to look at potential dealmakers and -breakers. After a summary of the main strategies and approaches under discussion for future or post-2012 mitigation efforts, including how the increasing problem of adaptation can be addressed, and the questions is raised what (continued) role existing instruments (i.e., the flexibility or project-based mechanisms) to reduce emissions could play.

Existing and new climate policy negotiation forums and mitigation approaches might be used to negotiate and decide on international and regional agreements and it will be interesting to see whether different emerging systems may be linked. Finally, after the examination of the potential inclusion or exclusion of certain key principles and criteria the question is asked whether there will be a gap post-2012 or not.

During the first quarter of 2007 EcoSecurities with the support of Oxford Climate Policy Group approached about 100 stakeholders, half of them have responded, on their views and expectations about whether and how climate mitigation efforts should or will continue. The interviewees ranged from climate policy experts,[1] non-governmental organizations (NGO), policy makers to private sector representatives.

2. Proposals on commitments and approaches for post-2012

Different proposals for a post-2012 regime addressing GHG mitigation as well as adaptation to climate change have been published and discussed so far. The strate-

---

* Principal Consultant, Policy Advisory and Capacity Building Services, EcoSecurities, email: robert@ecosecurities.com.
[1] This group consists of researchers, consultants and academics.

W.Th. Douma et al., eds., The Kyoto Protocol and Beyond
© 2007, T·M·C·Asser Press, *The Hague, The Netherlands and the Authors*

gies and approaches can be summarized in *quantified emission reduction targets with emission trading*, some of which are based on the basic architecture created by the Kyoto Protocol, and *non-emission target based approaches*, which articulate different types of commitments (see Figure 1). Binding emission targets can be based on the continuation of the currently adopted Kyoto-style absolute emission targets or based on flexible types of emission targets (e.g., dual targets, dynamic targets, positively binding targets, price caps, etc.). The main non-target based approaches proposed include proposals based on technology development and transfers, sectoral (i.e., non-target, but intensity based instead) agreements, policy based approaches, equity and development, as well as financial measures.

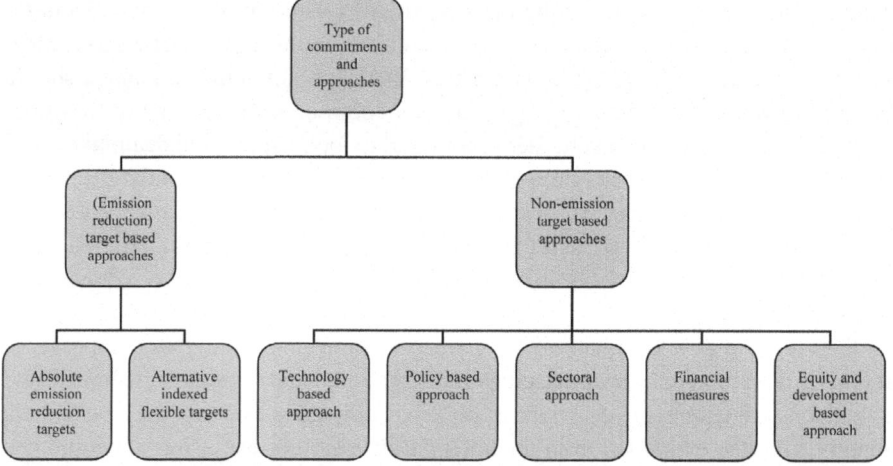

Figure 1. Overview of different types of possible future committments and approaches

Ninety percent of stakeholders interviewed support the continuation or expansion of absolute and binding emission reduction targets as it has been designed under the Kyoto Protocol. 33% of respondents that favor such a future regulatory framework support a mixed approach, i.e., shaping a consecutive climate change regime such that a target based approach exists parallel to voluntary or non-emission reduction target based approaches. They wish to involve as many participants as possible to reach the ultimate goal of the United Nations Framework Convention on Climate Change (UNFCCC) (see Figure 2 below).

Proposals under the group in favor of a mix of approaches also contain emission reduction target based approaches but are combined with non-emission reduction target based approaches such as technology or policy based approaches for non-Annex I and potentially also for Annex I countries. As shown in Figure 4, it can be seen that advantages for such a mixed approach are seen as manifold. Statements under the category *others* are, for example, that the system would be more acceptable to a wider stakeholder community while including different approaches. Other participants simply stated cost effectiveness. However, reaching the ultimate goal of the Convention, stabilizing GHG emissions to the atmosphere, and to find an

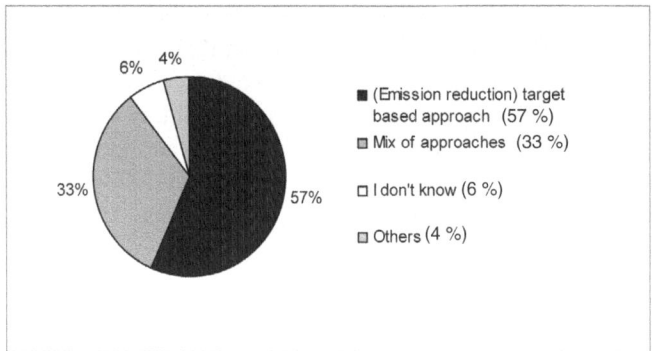

Figure 2. Most appropriate approach for a post-2012 international climate regime

agreement on (future) shared responsibilities between Annex I and non-Annex I countries or between Kyoto signatories and non-signatories is also deemed to be important.

The emission target based approach is preferred by a majority of policy makers, NGOs and also by the business sector, which can be explained due to its proven workability and pragmatic nature. Only a few participants stated that the mixed approach would be a simple solution. Efficiency and simplicity were not considered key criteria by the group that supports a mixed approach (Figure 3 and 4 below).

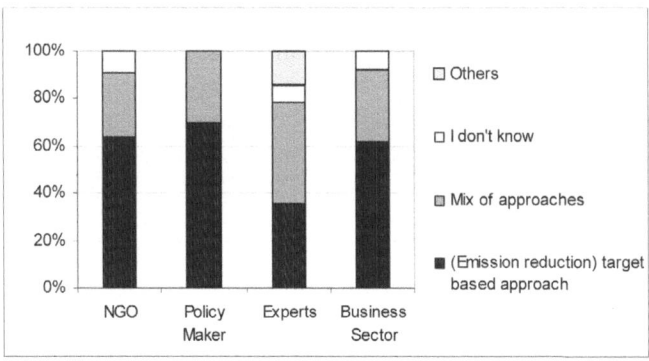

Figure 3. Proposed future approache(s) supported by which stakeholder group

An NGO representative stated that a *multistage approach* is probably the most likely approach that is going to be adopted. Such a system would allow for different types of commitments and progression towards absolute binding targets as countries develop. One main advantage would be that the creation of a totally new system from the scratch would not be needed by using existing and already implemented approaches, or at least elements, given that the timeline until 2012 is rather tight. The approach is perceived to be very pragmatic and has probably the most suitable

architecture. Over 20% of the interviewees who were in favor of an emission reduction target based approach mentioned that continuing or building on the Kyoto
architecture has a great advantage – it would make future negotiations simpler and
more efficient (see Figure 4 below). Continuing or building on Kyoto is seen as a
good way to reach the ultimate goal of the UNFCCC and to find and agreement
between Annex I and non-Annex I countries to the UNFCCC to continue and potentially even pave the way for non-Annex I commitments.

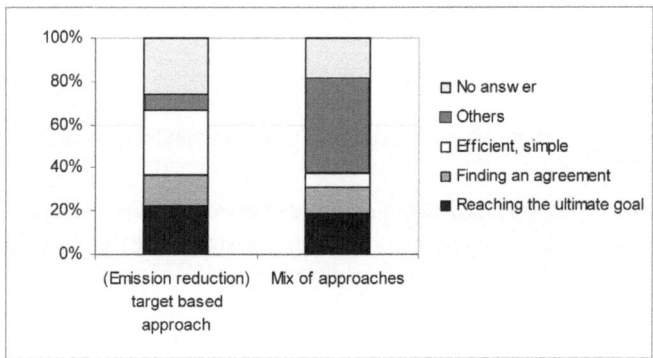

Figure 4. Perceived strengths and advantages of emission reduction target based approach
and mix of approaches

Overall, the absence of a clear perspective on what kind of international climate
regime will be in place after 2012 already represents a set of uncertainties and risks
for policy makers, different industrial sectors and most importantly, for the carbon-
finance sector. 15% of all interviewees highlight the necessity to find any kind of
agreement(s) whereas almost 20% voted for the importance of a simple and efficient system or to achieve the overall goal of mitigating emissions no matter how
this is achieved.

3. CONTINUATION OF PROJECT-BASED OR FLEXIBLE MECHANISMS SUCH AS
    THE CDM

Some proposals consider an extension or reform of the flexible mechanisms introduced by the Kyoto Protocol such as the CDM (Bodansky 2004) often referring to
the project-based character of these activities. With regard to the CDM, the main
aim of these proposals is to enable sectoral or policy-based activities to be considered under the CDM (either public or private sector initiated activities as the result
of a policy or a standard) that lower GHG emissions in a particular sector and
therefore generate Certified Emission Reductions (CER) (Sterk and Wittneben 2005
and van Schaik and Egenhofer 2005). In 2006 the CDM Executive Board already
released some preliminary guidance on a Programme of Activities (programmatic
CDM) which allows project activities under a programme of activities to be regis-

tered as a single CDM project activity, which means that a first step is taken to extend the CDM in its current form. Furthermore, uncertainties concerning the post-2012 climate framework and its implications for the continued existence of an international carbon market affect both the volume and the types of projects entering the CDM project pipeline. Conversely, if a clearer picture develops the flow of CERs from projects or programmes could increase quickly to fill the expected supply gap until 2012 and beyond – the EU already announced its continued and increasing commitment to climate change mitigation, including the CDM.

The lack of clear guidance on the role of the CDM after 2012, for example, is already affecting related investment decisions – certain project types and sectors are preferred due to short term carbon credit gains, the *low hanging fruits*. The vast majority of respondents see a continued role for CDM/JI post-2012. Though, 38% of out of the 90% respondents in favor of binding targets with a continued role for the flexibility mechanisms of the Kyoto Protocol to achieve those emphasize the need for reforms in the CDM. In addition, almost half of all interviewed stakeholders wishing sectoral or programmatic elements, almost a quarter of all interviewees highlighted the need for streamlined and simplified procedures and 17% of all respondents the need for considering sustainability aspects (Figure 5 and 6 below).

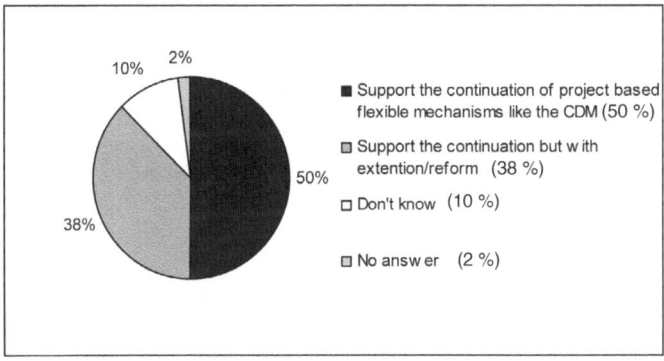

Figure 5. Continuation of project-based mechanisms post-2012

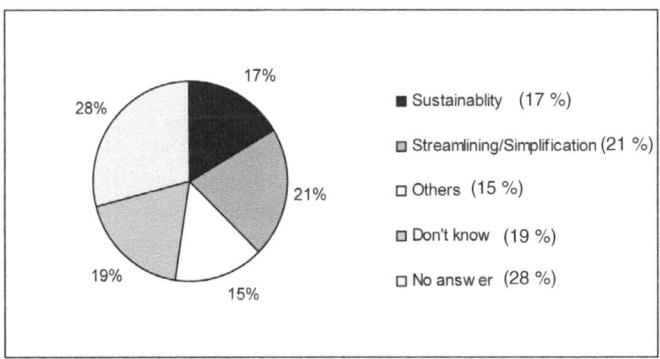

Figure 6. What reforms in the CDM?

Figure 7 and 8 show that the NGOs see a real need in reforming or extending the flexible mechanisms with the focus on sustainability aspects. The business sector is an ally here also calling for reforms or an extension of the CDM. However, in general the continuation of project-based mechanisms is perceived as very important by all stakeholder groups.

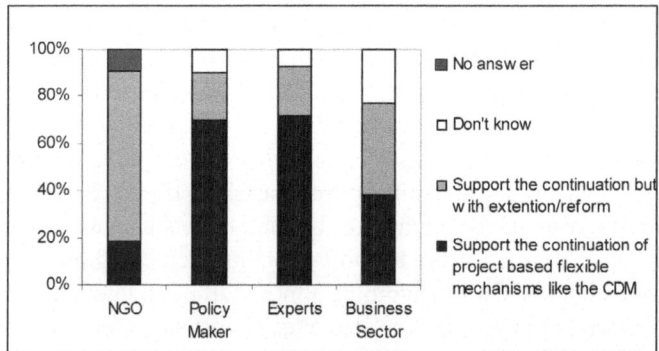

Figure 7. Continuation of the CDM and need for extension/reforms divided by stakeholder groups

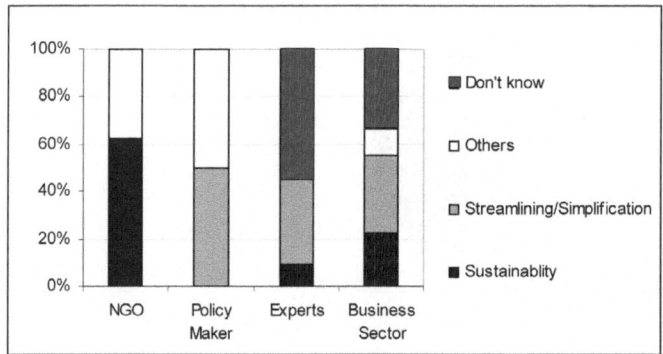

Figure 8. Important aspects regarding CDM reforms divided by stakeholder groups

4. ALTERNATIVE, NON-EMISSION REDUCTION TARGET BASED APPROACHES

A wide range of alternative non-emission reduction target based approaches have been proposed (Bodansky 2006), including non-target based but also target based approaches, partly seen as complementary to the Kyoto Protocol. The *Growth Baselines* and the *Converging Markets* proposals, for example, envision the possibility of sector based targets. *International Agreements on Energy Efficiency*, for instance, is a non-target based proposal where countries would develop energy efficiency standards. Technology based approaches such as *Technology-Centred Approach* propose the negotiation of protocols to finance collaborative research and develop common technology standards, and support deployment of existing and new technologies. A *Technology Backstop Protocol* would set forth medium - to long-term technology targets – for example, capture and sequestration of all carbon from new power plants by 2020 – in order to stimulate technology development and reduce emissions.

Many survey participants referred to two main approaches, the sectoral agreements and technology transfer based approaches. Technology (transfer) based approaches are seen to play a more important role with 64% of all respondents than sectoral agreements which are seen as a possibility and important by half of the interviewed stakeholders. However, particularly sectoral agreements are always seen as a parallel solution and not instead of approaches such as emission reduction targets (see Figure 9 and 10).

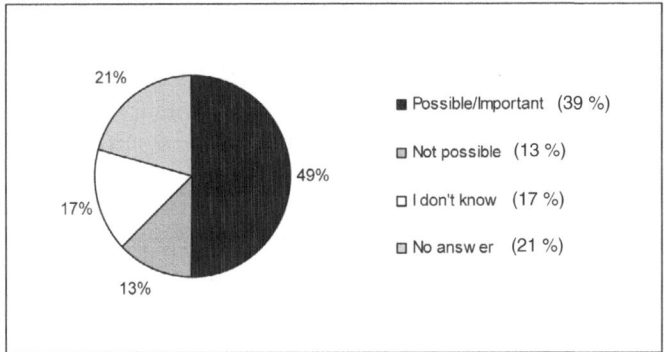

Figure 9. Inclusion of sectoral agreements post-2012

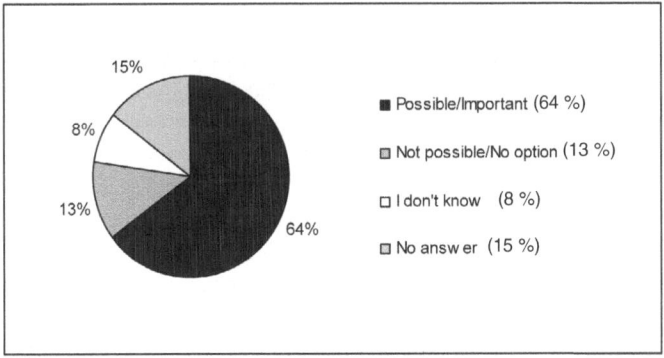

Figure 10. Inclusion of technology transfer based approaches post-2012

The survey confirmed that there is a group of stakeholders that support a *multidimensional* approach to effectively and fairly mitigate climate change, i.e., the negotiations of a post-2012 climate agreement should include multiple dimensions and mechanisms. Others argue that the negotiations about possible emission targets and allocation rights post-2012 have to be supplemented by other coalitions and commitments, under or even outside the umbrella of the UNFCCC. Parallel develop-

ments with different speeds is already happening and gaining support among those
that see no other way to effectively address climate change mitigation timely.

5.   NEGOTIATION ARENA

Some stakeholders believe that to effectively mitigate climate change, negotiations
and decisions on (a) future climate regime(s) may have to or will take place in
different arenas and actions may be conducted under different regulatory frameworks. Indeed, there are multiple dialogue forums on international action on climate change running in parallel with the negotiations under the UNFCCC.
Negotiations could, for example, take place more and more among G8 or G8+5
member states or major emitters and decisions being *transported* into the UNFCCC
afterwards. Such proceedings could potentially allow for the avoidance of certain
delays in decisions but might also avoid the sometimes crucial exchange of views
among parties to come to consensus based decisions with long-term effects. Some
stakeholders propose to move forward with negotiations outside the UNFCCC to
allow for groups that share a common understanding of how to tackle climate change
to move quicker and not being vetoed or veto others without any group really moving forward or rather implementing an approach effectively. This could or is already happening under the Asia-Pacific Partnership on Clean Development and
Climate[2] or under national or regional initiatives (i.e., RGGI[3]), for example, and
could become of increasing importance in the future.

More than 60% of the respondents see the UNFCCC still as the major negotiation forum, although a third of respondents believe negotiations should take place
in forums outside the UNFCCC (8%), or others influencing the UNFCCC (17%) or
wish to see the G8 or G8+5 as a (future) major driver (6%) (Figure 11). In reality,
less than half of the respondents see the UNFCCC and its negotiation and decision-
making bodies as the major forum (45%) whereas almost a third (29%) sees forums
outside the UNFCCC as the major constituencies and circles influencing the
UNFCCC. A mere 4% regard the G8/G8+5 or forums completely outside the
UNFCCC as major drivers with respect to action on climate change. Policy makers
see the UNFCCC as the most appropriate forum and framework where negotiations
should take place (Figure 12), interestingly enough only half of them think that this
is the place where negotiations will happen or decisions be made in reality (see
Figure 14). A similar picture appears in the NGO camp, although here more
interviewees see a continued important role for the UNFCCC in reality (see Figure
14). The business sector and climate policy experts desire and believe that different
forums influence the negotiations under the UNFCCC (Figure 13).

---

[2] The Asian-Pacific Partnership members (Australia, China, India, Japan, Republic of Korea
and the United States) have agreed to work together and with private sector partners to meet goals
for energy security, national air pollution reduction and climate change in ways that promote sustainable economic growth and poverty reduction. The partnership's focus is mainly on expanding investment and trade in cleaner energy technologies.

[3] <http://www.rggi.org/about.htm>.

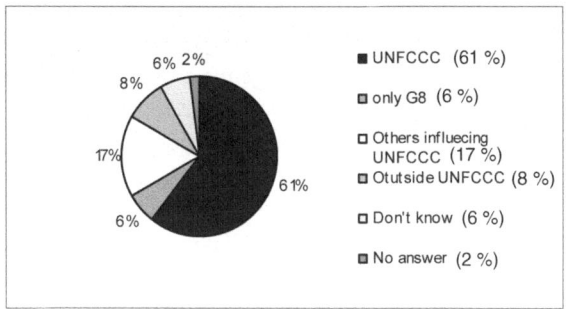

Figure 11. Forums in which the future climate change agreement should be negotiated

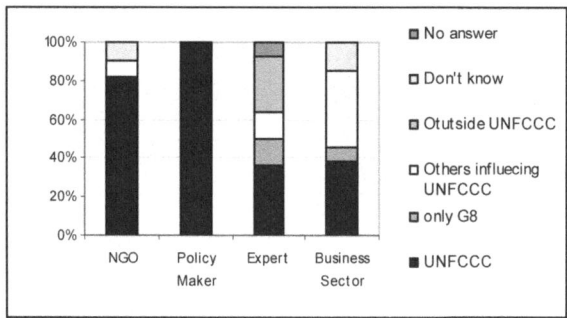

Figure 12. Desired negotiation forum divided by stakeholder group

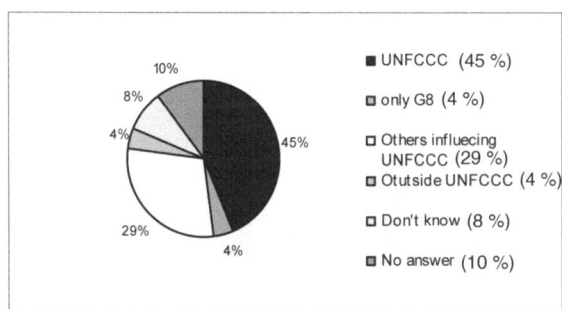

Figure 13. Forums in which a post-2012 climate agreement will be negotiated

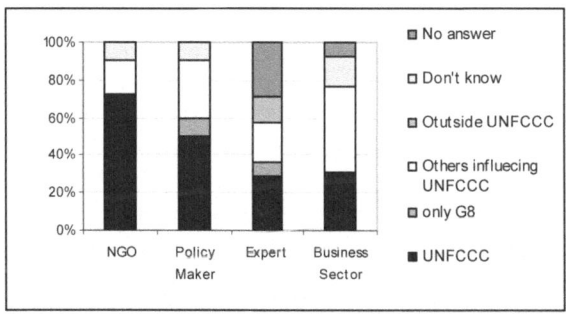

Figure 14. Forums in which negotiations will take place divided by stakeholder group

Almost half of all interviewees see the emergence of parallel climate change mitigation schemes (e.g., Asia-Pacific Partnership and other voluntary or non-binding initiatives) as desirable and acknowledge their increasing importance and role with respect to addressing climate change, compared to a quarter seeing this as no option at all (Figure 15). The policy makers are most in favor of parallel regimes because they create these schemes and certainly are in favor of the one they founded (see Figure 16). The expert community and the private sector see parallel regimes coming both with more than 40% and some NGOs are least open to such developments.

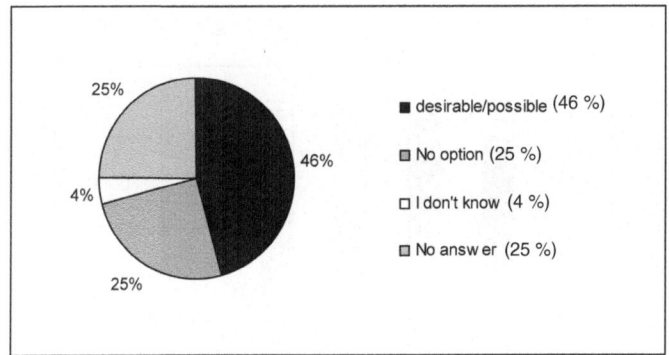

Figure 15. Occurrence of parallel climate policy regimes

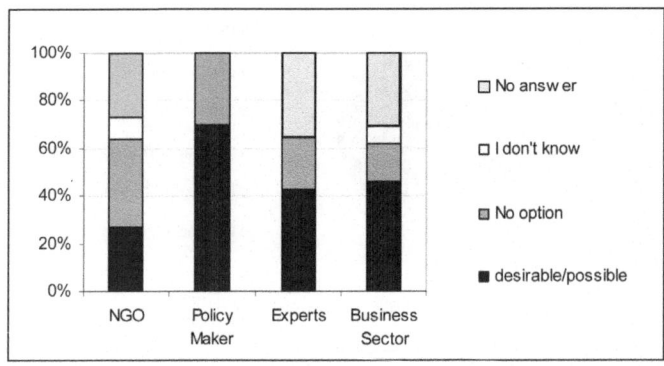

Figure 16. Occurrence of parallel climate policy regimes divided by stakeholders

6. LINKING OF DIFFERENT EMISSION TRADING SCHEMES

Currently, the international carbon market is dominated by the EU Emissions Trading Scheme (EU ETS) which started in 2005, a mandatory scheme that also assists the EU and its member states to meet their targets under the Kyoto Protocol. Compliance or trading units are EU allowances (EUA) but thanks to the Linking Directive CERs and ERUs from the Kyoto Protocol's project-based mechanisms can be used within the EU ETS from 2005 and 2008, respectively. The EU ETS is a policy tool for managing emissions of firms in key industrial sectors. The EU ETS is based on a recognition that creating a price for carbon through the establishment of a liquid market for emission reductions offers the most cost-effective way for EU member states to meet their Kyoto obligations and move towards the low-carbon economy of the future.

Emissions trading systems of various forms are being implemented by a number of countries inside and outside the signatories of the Kyoto Protocol. These trading schemes are part of the strategies of countries, states or companies to address climate change and contribute to climate change mitigation, such as the New South Wales Abatement Scheme, the Canadian Emission Trading Scheme, the California Climate Action Registry, the Chicago Climate Exchange, the Japan Voluntary Emission Trading Scheme or the RGGI. Theoretically, emissions trading schemes with similar regulatory frameworks could be linked in the future or under a new post-2012 umbrella regime. Although the detailed regulatory requirements and legal implications of such undertakings would need to be fleshed out – the EU ETS, for example, through its Linking Directive opens the door to other trading schemes. In this context the question arises whether existing mechanisms such as the CDM or other project based offsetting mechanisms, using the same currency (i.e., tCO_2), can play a role in linking different systems.

The emergence of national, sub-national or regional emission trading schemes has raised the issue of linking those schemes to achieve more ambitious and balanced efforts for GHG emission reductions whilst also promoting economic efficiency through the use of similar instruments and a common currency, i.e., tCO_2. 80% of all participants in the survey consider the linking of other (sub-)national or regional trading systems with the EU ETS as desirable and possible (Figure 17). Almost all policy makers (90%) see this opportunity as desirable and possible followed by the NGO community with more than 80%. Among the experts and the business sector more or less 20% see the linking as not desirable as well as not possible.

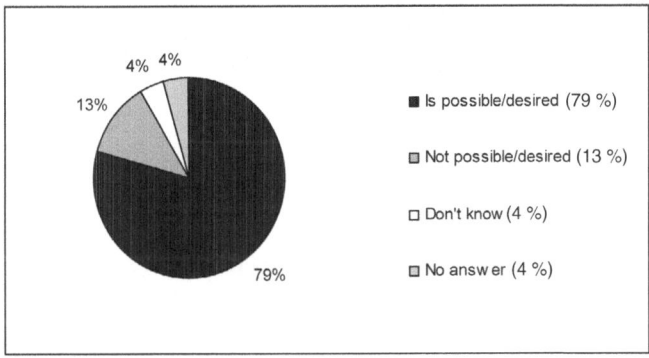

Figure 17. Linking different emission trading schemes

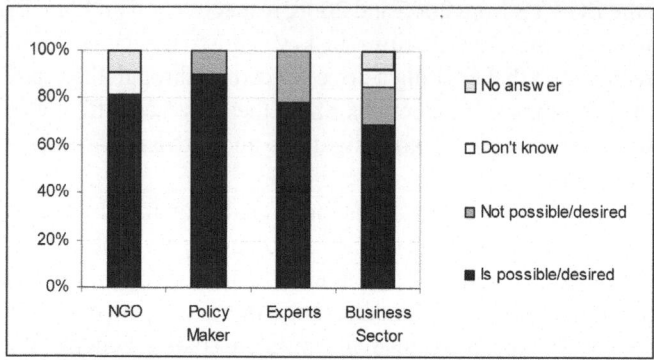

Figure 18. Linking different emission trading schemes divided by stakeholder groups

Figure 19 shows that almost half (46%) of all interviewees could see a role for project based mechanisms such as the CDM or similar mechanisms here. This would certainly have technical and legal consequences to be solved due to the EU ETS being based on a compliance system with binding targets whereas some of the other trading systems only share certain elements with the EU ETS, i.e., flexibility or offsetting mechanisms. However, it is also to be considered that almost 50% of the interviewees could not or would not answer the question which underlines the question of how the linking should or could be done technically. And to what extent existing tools and mechanisms could be used or would need to be amended or even completely new ones would be needed.

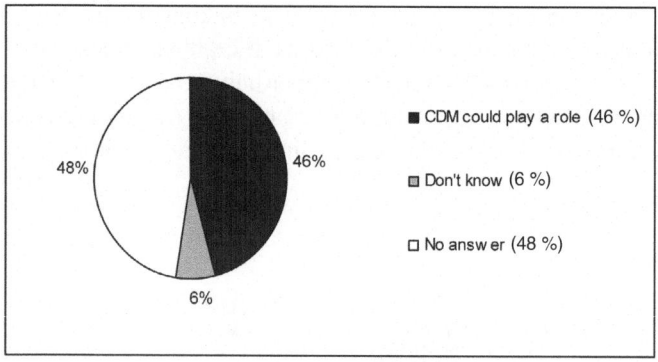

Figure 19. A role for project based or flexible mechanisms in linking different schemes?

7. GUIDING PRINCIPLES OF A FUTURE CLIMATE REGIME

A whole range of key principles exist which, in theory, should actually guide the design of a post-2012 international climate regime. Based on experiences from negotiating the rules and regulations of the Kyoto Protocol and its mechanisms it is also interesting to ask for the perceptions of the different stakeholder groups regarding the desired principles and the driving principles in reality (see Figure 20 and 21).

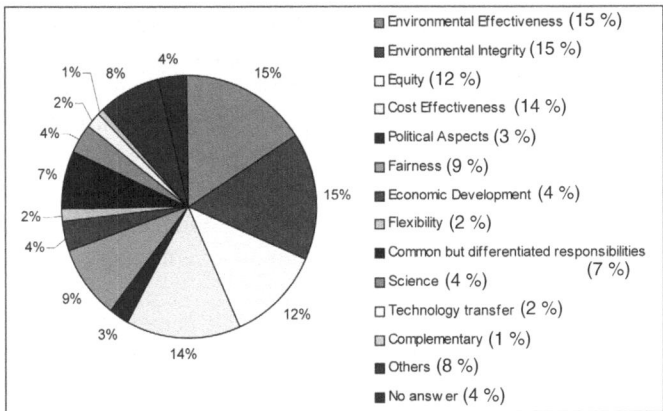

Figure 20. Which should be the guiding principles behind a post-2012 climate agreement?

As shown in Figure 20 environmental effectiveness (15%), environmental integrity (15%), cost effectiveness (14%), equity (12%) are the desired key principles which were named most often. Compared with the principles which respondents think will guide the negotiations on a future climate change mitigation regime the lion share hold cost effectiveness (15%) followed by others (13%), political aspects (12%) and economic development (10%). So in reality rather economical and political aspects will influence the decisions around such an agreement instead of the environment related principles.

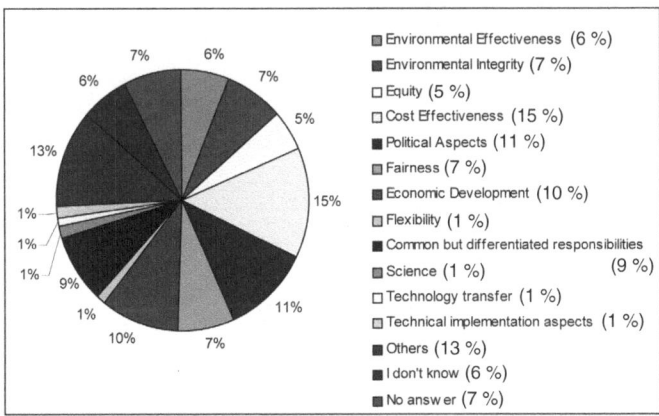

Figure 21. Which will be the guiding principles behind a post-2012 climate agreement?

8. A GAP AFTER POST-2012?

Both climate policy and carbon market specialists have emphasized that to secure continued and future investments in clean energy projects, contributing to climate change mitigation, it is important that the market and investors have early signals about a future climate regime. This is particularly important with regard to the future of project based mechanisms like the CDM. The future of the CDM and investments in clean energy projects has to be seen on the background of the question whether there will be a continuation of the Kyoto Protocol or rather a second commitment period or whether a new climate mitigation regime will take over post-2012. Finally, the possibility of a gap without binding emission targets or any other international regime in place is also existent.

Figure 22 and 23 demonstrate a positive overall perception with respect to a continuation without a gap between a second commitment period of the Kyoto Protocol or (a) future climate change mitigation agreement(s) (64%). The vast majority of the participants also highlighted the importance to find an agreement in time. Representatives from the business sector were most pessimistic about this question and most negative answers came from this group. Policy makers and NGO representatives were the most positive groups about having no gap post-2012 with almost 90% and more than 70% respectively.

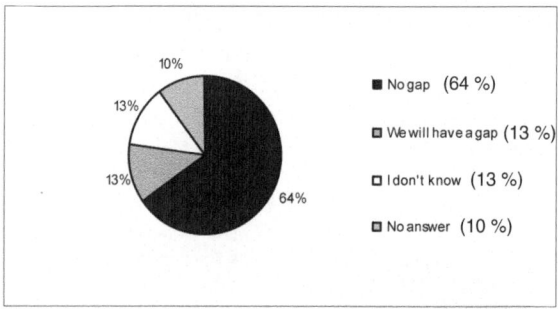

Figure 22. A gap post-2012 or a continuation of the Kyoto Protocol or a new climate change mitigation regime?

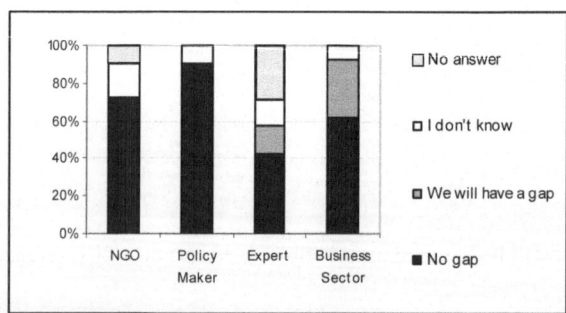

Figure 23. A gap post-2012 or a continuation of the Kyoto Protocol or a new climate change mitigation regime (divided by stakeholders)?

9. CONCLUSIONS

This survey provides insights on stakeholders' opinion from four different groups (i.e., policy makers, experts, NGOs and business sector) about the future of the international climate change mitigation regime and the carbon market. It sheds light on stakeholder groups' views about related issues such as what kinds of commitments will be in place in the future or post-2012, i.e., emission target based or other approaches, whether there is a continued role for flexible or project-based mechanisms and whether these mechanisms can be used to link up different emission trading schemes. Finally, the questions of what kind of guiding principles will play a role during the coming negotiations and whether there will be a gap post-2012 have been investigated.

Almost all stakeholders across all stakeholder groups support the continuation of approaches based on emission reduction targets, although a third are in favor of the inclusion of approaches not based on emission reduction targets under a new or post-2012 international climate change agreement. Highlighting the parallel development and implementation of these approaches, mostly sectoral and technology based approaches are mentioned with technology or technology transfer based approaches having a slightly higher share – this might be related to the clearer link to promotion and deployment of clean or energy efficiency technologies and renewable energy targets, at least some stakeholders seem to think in this direction. One out of five survey participants stressed the importance of achieving the ultimate goal of the UNFCCC through any approach or mix of approaches post-2012, as well as a simple and efficient implementation framework.

The discussions on the continuation or a second commitment period of the Kyoto Protocol or what might come after the Kyoto Protocol and might replace it seriously started after the Kyoto Protocol entered into force in 2005. This also begins to affect the development of projects under the CDM, largely a successful mechanism for the implementation of the Protocol, due to the problem of the window of opportunities closing soon and the long(er) lead times for some project types. On the background of the success of the CDM and the announcement of the EU to continue the EU ETS, including the CDM, also the large majority of stakeholders support a continuation of the CDM with more than a third seeing the need for reforms. Particularly, the two groups rather concerned about the implementation, the business sector and NGOs, highlight the importance of streamlining and simplifying the mechanism as well as reintroducing sustainability aspects.

The potential linking of regional and national or sub-national emission trading systems with the EU ETS, being the dominant system at the moment with and open to linking up with other schemes, is supported by almost 80% of the survey participants and by all stakeholder groups. Mainly experts and the business community raise some concerns about the possibility of doing so or whether this is desirable at all. These concerns are related to the obvious technicalities and legal issues that such undertakings would entail.

The survey revealed that also for any kind of new climate change regime or agreement economic and political aspects will be the main drivers again behind the negotiations as they have been high on the agenda during the negotiations about the rules and regulations for the implementation of the Kyoto Protocol. Environmental effectiveness and integrity score highest when asked for the principles that should guide the negotiations whereas cost-effectiveness and political motives are mentioned when it comes to the realities of negotiations.

Similar the majority of respondents see the UNFCCC as the appropriate framework where negotiations should take place but in reality less than half think that negotiations will solely take place or even decisions will be made under the UNFCCC framework. Almost a third of all interviews see circles outside the UNFCCC, the importance of the G8/G8+5 is stressed by some, increasingly influencing the negotiations or rather making decisions beforehand that will then be transported into the UNFCCC then.

46% of all survey participants consider the emergence of parallel climate change mitigation regimes such the Asia-Pacific Partnership and other voluntary or non-binding initiatives as a desirable and important development. The group of policy makers leads with respect to supporting these developments, followed by experts and the private sector with NGOs being less in favor of the emergence of parallel regimes. Policy makers are rather in favor of more options, in particular when they offer to potentially buy in countries or country groups that have so far rejected serious involvement in global mitigation efforts. The NGO community often sees such initiatives as a cheap excuse to not signing up to serious and binding commitments. These two groups also raise the most concerns with respect to parallel climate change mitigation regimes. Experts and the business sector mostly support such developments with only some resistance.

With 64% the majority of interviewees believe there will be no gap between the end of the first commitment period of the Kyoto Protocol and the second commitment period or a future climate regime. 13%, however, believe there will be a gap post-2012. Here the most pessimistic views come from the experts, policy makers and NGO representatives are most positive about having no gap post-2012.

For project developers or suppliers, as well as carbon credit buyers with a focus on the Kyoto Protocol compliance market and the EU ETS the survey results reflect three important messages: 1. The majority supports a continuation of Kyoto-style absolute emission targets. 2. The CDM is seen as a crucial element of such a continuation of an emissions reduction scheme with binding targets. 3. Two thirds of all interviewees believe there will be no gap between the completion of the first commitment period of the Kyoto Protocol and a second commitment period or another future climate change regime. However, it will be interesting to see whether at all and how the international climate community will respond to the various other, alternative mitigation routes that have emerged or emerge, be they voluntary or of other non-binding nature with respect to emission reductions. The UNFCCC should be and provide the overarching umbrella for all kinds of agreements but it will be a challenge to pool the different processes and speeds of these processes,

partially emerging outside the UNFCCC negotiation forum, that are less or differently regulated and controlled as the Kyoto Protocol compliant activities.

References

Blok, K., Hoehne, N., Torvanger, A. and Janzic, R. (2005): 'Towards a Post-2012 Climate
    Change Regime', Final Report, Contracted by European Commission, DG Environment,
    Directorate C – Air Quality, Climate Change, Chemicals and Biotechnology.
Bodansky D. (2004): 'International Climate efforts beyond 2012: A survey of approaches'.
    Prepared for the Pew Center on Global Climate Change
Den Elzen, M., Berk, M., Lucas, P., Criqui, P. and Kitous, A. (2006): 'Multi-Stage: A Rule-
    Based Evolution of Future Commitments under the Climate Change Convention', Inter-
    national Environmental Agreements No. 6, 1-28.
Hoehne, N., Phylipsen, D. and Moltmann, S. (2006): 'Factors underpinning future action',
    EcoFys for the Department for Environmental Food and Rural Affairs (DEFRA), UK.
Michaelowa, A. (2006): 'Principles of Climate Policy after 2012', *Intereconomics*, March/
    April 2006.
Pointcarbon (2007): 'Rich countries should take on ambitious targets this year', UNFCCC
    chief says. Downloaded on 15 March, 2007, from: <http://www.pointcarbon.com/article.
    php?articleID=20856>.
Van Schaik, L. and Egenhofer, C. (2005): 'The State of the CDM Debate'. European Cli-
    mate Platform, Background Paper No. 1.
Sterk, W. and Wittneben, B. (2005): 'Addressing Opportunities and Challenges of a Sectoral
    Approach to the Clean Development Mechanism', *JIKO Policy Paper* 1/2005.
Stuart, M., Ratton, M., and Tippmann R. (2006): 'Deja vue All Over Again – Considering
    the post-2012 era', Carbon Market Monitor, Point Carbon.
Tippmann, R. (2007): 'Stakeholder views on future climate change mitigation efforts'. *Joint
    Implementation Quarterly*, Vol. 13 – No. 1.
UNDP (2006): 'The Clean Development Mechanism – An Assessment of Progress', No-
    vember 2006.

THE KYOTO PROTOCOL AS A PIONEER AMONG THE MULTILATERAL ENVIRONMENTAL AGREEMENTS

Michael Bothe*

Why is the Kyoto Protocol (KP) currently at the cutting edge of international law? The first distinguishing element is its scope. Like other multilateral environmental agreements (MEAs), it addresses a particularly well-defined environmental problem. But in contradistinction to other MEAs, it is a problem which, if not solved, will fundamentally affect the living conditions of humankind, probably anywhere in the world. Furthermore, and also in contradistinction to other MEAs, the measures taken to solve the problem also comprehensively affect lifestyle and economic conditions in the participating States.

The second distinguishing element is the innovative policy mix pursued by the KP. It uses a number of regulatory techniques which are also found in other recent MEAs, but it adds new ones which have not been tested at the international level. The broad scope of the KP, in conjunction with this ingenious policy mix, entails an unprecedented challenge for an international administration. It is a bold attempt to solve a global problem through an administrative mechanism of global scope.

1. THE PROBLEM AND ITS REMEDIES

The issue becomes clear if one recalls the whole array of measures to be taken to combat climate change:

Leaving aside the option of living with climate change by way of adaptation, there are, basically, two ways to combat climate change:

- Reduce greenhouse gas (CO_2, methane, ozone etc.) emissions;
- Promote greenhouse gas sequestration.

How can this be achieved? There are a number of ways:

a) Reduce emissions, for example CO_2,
  - by reducing fossil fuel combustion by fixed or mobile sources:

---

* Professor em. University of Frankfurt/Main.

W.Th. Douma et al., eds., The Kyoto Protocol and Beyond
© *2007, T·M·C·ASSER PRESS, The Hague, The Netherlands and the Authors*

- fixed sources:
  - invest in alternative energy production methods (nuclear, water, wind);
  - increase the efficiency of industrial installations;
  - decrease energy demand by increasing energy efficiency or changing energy consumption habits or needs;
- mobile sources:
  - use alternative types of engines or fuels;
  - speed limits or other limitations on the use of cars;
- by capturing CO_2 emissions.

b) Sequestration of greenhouse gases through:
   Afforestation and reforestation – which entail far-reaching land use and land use change decisions.

Each of these measures presents problems of its own. Drawing an environmental balance sheet is thus required. There are specific drawbacks involved in most of these measures. Nuclear energy does not produce CO_2 emissions, but has definite other environmental disadvantages. The use of bio-fuels avoids using fossil fuels, but their production uses a great deal of valuable space and leads to monocultures, which are problematic from an environmental point of view. There is a similar tension between afforestation and biodiversity.

2.   THE REGULATORY CHALLENGES

2.1   **The comprehensive scope of measures**

The first challenge is the scope of measures to be taken for the implementation of the KP.
   The broad array of measures just described fundamentally affects choices of national economic policy and the lifestyle of citizens, more than any other global regulatory regime does. The overall economic impact of these measures (or of abstaining from taking these measures) is enormous.
   Global regimes addressing the use of the global commons and their resources have a long tradition. But they relate just to one type of commons: the seas. They thus affect a limited (though important) part of human and societal interests:

- food production (fisheries);
- trade (shipping);
- production of other, in particular mineral resources (a relatively recent issue);
- security (naval warfare).

The interests affected by the implementation of the KP are much broader, they are really comprehensive. The social, political and economic impacts of some of these measures affect in an unprecedented way many fields of foreign and internal policy. This is the social and political challenge of the KP system.

2.2 Managing uncertainty

The second challenge: regulation on the basis of the precautionary principle. The point of departure of the earlier negotiations was a prognostication involving a high degree of uncertainty as to the risks of climate change. That uncertainty has been decreasing, but uncertainties remain. Uncertainty means that the political obstacles which measures to protect the environment have to overcome are greater than in the case where the danger is clear and immediate (for example, oil spills). Such obstacles are difficult to overcome at the national level, but even more so at the international level. It is one of the political achievements of the KP that a far-reaching regulation has been reached despite the remaining uncertainties on the basis of the precautionary principle. Yet, here lies the root of the American objections.

2.3 Managing complexity

It has been shown that there are many chains of causation which may contribute to the desired result of slowing down the increase of the greenhouse effect. The regulatory problem, in a technical perspective, is to design and implement a normative framework which influences these chains of causation in a way which leads to the effective solution of the climate problem. This raises the question of the regulatory instruments to be used. It is in this area that the KP has brought major innovations.

3.   THE REGULATORY INSTRUMENTS

3.1 Regulatory instruments in a multilevel system

The basic regulatory tool adopted by the KP is a limitation of the net aggregate emissions of the greenhouse gases of each participating country – a regulatory instrument which is not uncontroversial. The U.S. proposes a completely different approach. And indeed, there are different ways in which regulation can address these various chains of causation.

As to aggregate standards, one has to distinguish between emission standards and quality standards. The technique of using an aggregate quality standard as a yardstick for international obligations is not uncommon. For example, water of a certain quality has to be delivered at the border point. The first treaty to use aggregate national emissions as standard for combating air pollution was the 1985 Helsinki Protocol on SO_2 emissions, a protocol additional to LRTAP. The advantage of this approach: it fixes a relevant environmental goal and leaves the States with a large degree of freedom as to how to reach it. In the case of the Helsinki Protocol, the parties thereto are free to use nuclear energy to be able to fulfil their SO_2 reduction obligations. A similar but more sophisticated approach is that of the Montreal Protocol: an aggregate production and consumption limitation of certain substances as a means to reduce their emissions.

KP uses this very regulatory approach by defining the obligation in terms of quantified emission limitation and reduction commitment (QELRC), defined, first, as a percentage of an existing quantity of emissions (overall goal: reduction by 5% in relation to the 1990 level of emissions), which is then translated into "assigned amounts". In this respect, it is based on the recent precedents just described. But it is much more complex.

First, it addresses an array of six substances which have to be brought together as one aggregate standard: the "anthropogenic carbon dioxide equivalent", i.e. the standard is based on a calculation which equals the greenhouse effect of a quantity of a particular greenhouse gas to that of a unit of carbon dioxide. This calculation is an exercise in physics.

Second, it is based on one single goal (a reduction by 5% of the aggregate emissions of all participating States), but this goal is to be achieved through unequal obligations depending on the different situation of the participating countries (between 8% decrease to a limitation of an increase of up to 10%), in contradistinction to the uniform 30% reduction obligation under the 1985 Helsinki Protocol. This differentiation is not calculated, it is negotiated!

The third element is the fact that the standards are net emissions. Sequestration of greenhouse gases, in particular in vegetation, is deducted from a country's emissions. As the exact amount of sequestration is difficult to determine, there must be agreement on the way of calculating sequestration.

All three elements of complexity entail consequences at the level of implementation.

3.2 Flexible mechanisms and "economic instruments"

To address these problems in an economically reasonable way, the KP uses a number of so-called "flexible mechanisms" – which add to the complexity of the system. In view of the high cost of the measures to be taken to redress the situation, these measures must be both effective and efficient. A major way to ensure economic efficiency is the systematic use of flexible mechanisms, allowing for emission reductions or sequestration of greenhouse gases being achieved at those places where they can be achieved at the lowest cost. The basic principle underlying the four types of flexibilisation is the same: a lower reduction achieved in one place is compensated by a higher reduction made elsewhere, where it is more appropriate, in particular less expensive, to do so. This means that the overall beneficial effect for the environment remains the same, but is achieved at a lesser price. Because these mechanisms at least in part rely on a market mechanism, they are called "economic instruments". This is to a certain extent a misnomer, as these instruments require a high degree of sophisticated regulation in order to function.

At the national level, there exist certain experiences in respect of these instruments, in particular in the United States. At the international level, the systematic use of these flexible instruments is unprecedented. They are, so to say, the essential trademark of the Kyoto system.

There are four such mechanisms:

- the "bubble" (Art. 4 KP);
- Joint Implementation (JI, Art. 6);

- The Clean Development Mechanism (CDM, Art. 12);
- Emission trading (Art. 17).

The transfer process involved in each of these mechanisms is different. Art. 4: the bubble allows a group of States to redistribute there limitation or reduction obligations provided that their combined aggregate reduction remains the same. This possibility has been used by the EC, and this is commonly called the "EU Bubble". JI and CDM have in common that they are related to projects. The reduction achieved through a greenhouse gas reduction or sequestration project in one place is credited to a limitation or reduction obligation of another country. It is understood that this transfer of QELRs is accompanied by a reverse cash flow. Emission trading, too, constitutes an offset between, on the one hand, emission reductions achieved in one country in excess of the country's obligations and, on the other hand, reductions not reached in another, that transfer also being accompanied by a reverse transfer of cash.

It goes without saying that these instruments, in order to function, require precise bookkeeping. Especially the project-related mechanisms require a precise evaluation of the effects of each project in terms of emission reduction or greenhouse gas sequestration and a documentation of that transaction.

This is a crucial element involved in the KP system.

4. ENSURING COMPLIANCE – INTERNATIONAL ADMINISTRATION

These complex mechanisms will only function to the satisfaction of all parties to the regime if there is a reasonably reliable system of ensuring compliance. The high economic impact of the measures does not tolerate windfall profits. This is the purpose of the institutional apparatus created by the KP and the decisions of the COP/MOP. It is very elaborate, more elaborate than that of any other universal regulatory regime. That is the final reason why the KP is a pioneering regime in international relations, whatever its remaining deficiencies may be.

As already said, the system requires accurate bookkeeping due to the double complexity of the system: the basic QELRCs are differentiated (different for different entities) and composite (composed of a variety of greenhouse gases and their sequestration), and their calculation is constantly modified by the use of the flexible mechanisms. This is the distinct challenge of the Kyoto system of ensuring compliance. Never has any international administration tried to cope with so complex a matter – complex both from a political and a technical point of view.

True, there exist precedents for procedures to ensure compliance with MEAs. The creative moment were the deliberations conducted under the Montreal Protocol in order to create a "compliance mechanism" in the early 1990s. This apparently corresponded with a basic need of the MEA regimes. It has triggered a development of such mechanisms for most major MEAs. But it is the compliance system created under the KP which, due to the complex character of the obligations to be complied with, excels in complexity. This is shown by the sheer amount of norms which govern the implementation process under the KP, contained in the Marrakech Accords.

The basic structures of these procedures are similar for all MEAs which contain this type of procedure. There is a first phase of collecting information. Its starting point is self-reporting by States. In order to be meaningful, they must be comparable – and this requires standard-setting by the regime.

The next phase is the evaluation of the reported data as to their reliability. An assessment of compliance can only be made on the basis of correct data. If that assessment leads to a doubt as to compliance, the issue can be brought before the "compliance mechanism" *stricto sensu*. It has, and this has become the usual approach, two aspects, in the case of the KP two branches: a facilitative branch and an enforcement branch. The former will help the State to rediscover the way of compliance, the latter may state that a violation has occurred and decide on negative sanctions.

There are two types of sanctions depending on the type of obligation – and this, too, is an important innovation: sanctions for the violation of procedural duties (monitoring and reporting) are clear and strict: the defaulting State loses the right to participate in the flexible mechanisms. As that participation may be a kind of emergency kit to finally achieve compliance, this sanction is highly relevant. The sanction for failing to achieve the applicable QELR is a modest increase in the debited amount which is carried over to the next reduction period. Whether and to what extent this is a real threat remains to be seen when the type of obligations imposed in the next reduction period becomes clear.

Thus, as far as implementation is concerned, the KP relies heavily, on the one hand, on recent innovations of the international system, in particular in recent additions to MEA regimes. Its complexity provides the real test of the viability of these innovations. But, in addition, the KP provides for a number of innovations of its own. All this accounts for its pioneering character.

5. CLIMATE CHANGE UNDER DOUBLE JEOPARDY: TECHNICITY AND UNIVERSALITY

The climate change regime is under a special kind of double jeopardy: On the one hand, it has to develop an unprecedented regime of international administration – the test of its ability to do so is just starting. On the other hand, the economic development of States which are not parties to the KP or do not have QELRC makes it questionable whether a meaningful cap on the worldwide emission of greenhouse gases can be achieved even if the parties to the KP attain their stated goal.

The road ahead is rough, but it is the specialty of pioneers that they advance on difficult terrain.